McGraw-Hill's

500

College Chemistry

Questions

D1445982

Also in McGraw-Hill's 500 Questions Series

McGraw-Hill's

500

College Chemistry

Questions

Ace your College Exams

David E. Goldberg, PhD

New York   Chicago   San Francisco   Lisbon   London   Madrid   Mexico City
Milan   New Delhi   San Juan   Seoul   Singapore   Sydney   Toronto

The McGraw·Hill Companies

David E. Goldberg, PhD, was formerly professor of chemistry and department chairman at Brooklyn College City University of New York. He is the author of widely used chemistry textbooks as well as *Schaum's Outline of Beginning Chemistry.*

1 2 3 4 5 6 7 8 9 10 11 12 13 14 15 16 17 18   QFR/QFR   1 9 8 7 6 5 4 3 2

ISBN   978-0-07-179700-9
MHID   0-07-179700-9

e-ISBN   978-0-07-179701-6
e-MHID   0-07-179701-7

Library of Congress Control Number 2011944581

McGraw-Hill products are available at special quantity discounts to use as premiums and sales promotions or for use in corporate training programs. To contact a representative, please e-mail us at bulksales@mcgraw-hill.com.

This book is printed on acid-free paper.

CONTENTS

INTRODUCTION

You've taken a big step toward success in chemistry by purchasing *McGraw-Hill's 500 College Chemistry Questions*. We are here to help you take the next step and score high on your first-year exams!

This book gives you 500 exam-style questions that cover all the most essential course material. Each question is clearly explained in the answer key. The questions will give you valuable independent practice to supplement your regular textbook and the ground you have already covered in your class.

This book and the others in the series were written by experienced teachers who know the subject inside and out and can indentify crucial information as well as the kinds of questions that are most likely to appear on exams.

You might be the kind of student who needs extra study before the exam for a final review. Or you might be the kind of student who puts off preparing until the last minute before the test. No matter what your preparation style, you will benefit from reviewing these 500 questions, which closely parallel the content and degree of difficulty of the questions on actual exams. These questions and the explanations in the answer key are the ideal last-minute study tool.

If you practice with all the questions and answers in this book, we are certain you will build the skills and confidence needed to excel on your exams.

—Editors of McGraw-Hill Education

CHAPTER 1

Measurement

Metric System

1. Evaluate: $\dfrac{6 \times 10^7 \text{ cm}^3}{(2 \times 10^4 \text{ cm})(3 \times 10^3 \text{ cm})}$

2. (A) Express 4.23 m in kilometers, in centimeters, and in millimeters.
   (B) Express 29.66 mm in centimeters and in meters.

3. (A) How many cubic centimeters are in 1 m^3?
   (B) How many liters are in 1 m^3?
   (C) How many cubic centimeters are in 1 L?

4. Find the capacity in liters of a tank 0.50 m long, 20 cm wide, and 25 mm deep.

5. Convert
   (A) 3.00×10^2 g to milligrams
   (B) 1.90×10^{-2} cm^3 to liters
   (C) 6.21 km to millimeters
   (D) 3.33 mL to cubic centimeters
   (E) 9.70×10^6 mg to kilograms
   (F) 2.22×10^{-3} L to milliliters
   (G) 6.55 g/L to grams per milliliter
   (H) 4.18 kg/L to grams per milliliter

6. Perform the following calculations.
   (A) $(3.00 \times 10^2 \text{ cm}) + (7.50 \times 10^{-1} \text{ m})$
   (B) $(0.200 \text{ m}) (1.00 \times 10^{-4} \text{ km}) (3.00 \times 10^2 \text{ mm})$

Significant Figures

7. How many significant figures are in each of the following numbers?
   (A) 23
   (B) 119
   (C) 2.170
   (D) 0.00010
   (E) 1.00×10^6
   (F) 1000
   (G) π

8. Perform the following operations:

   (A)
   $$\begin{array}{r} 73.01 \text{ cm} \\ 17.3 \ \text{ cm} \\ +\ 0.11 \text{ cm} \\ \hline \end{array}$$

   (B)
   $$\begin{array}{r} 198 \ \text{ g} \\ -\quad 2.2 \text{ g} \\ \hline \end{array}$$

9. (A) $12.7 \times 11.2 = ?$
   (B) $108/7.1 = ?$

10. Underline each significant digit in the numbers below. If the digit is uncertain, place a question mark below it.
    (A) 1.066
    (B) 750
    (C) 0.050
    (D) 0.6070
    (E) 40.0

11. Calculate to the proper number of significant digits:
    (A) $(4.50 \times 10^2 \text{ m}) + (3.00 \times 10^6 \text{ mm})$
    (B) $(4.50 \times 10^2 \text{ cm}) (2.00 \times 10^6 \text{ cm})$
    (C) $(4.50 \times 10^2 \text{ mL}) - (0.0225 \text{ L})$

12. Consider three solutions with the following hydronium ion concentrations (to be discussed in Chapter 20):

$$1.99 \times 10^{-2} \text{ M} \qquad 1.99 \times 10^{-9} \text{ M} \qquad 1.99 \times 10^{-12} \text{ M}$$

(A) How many significant digits are in each concentration?
(B) What logarithm (log) value is shown on a simple scientific calculator for each?
(C) Which digits are the significant digits in each calculator value?

13. Compute and state the answers in standard exponential form.

(A) $(2.0 \times 10^{13} \text{ cm}) + (1.5 \times 10^{14} \text{ cm})$
(B) $(8.0 \times 10^{-14} \text{ cm}^2)/(4.0 \times 10^{-13} \text{ cm})$
(C) $(5.0 \times 10^{17} \text{ cm})(2.0 \times 10^{-4} \text{ cm}^2)$
(D) $(6.6 \times 10^{15} \text{ cm}) - (3.0 \times 10^{16} \text{ cm})$

CHAPTER 2

Structure of Matter

Elements, Compounds, Mixtures

14. A certain combination of iron and sulfur is partially soluble in excess CS_2. The combination may be described as a _____.

15. A certain homogeneous material has a melting point of 73.7°C. When 10.0 g of the material is placed in 20 mL of water, only 2.0 g of the material dissolves. When the remaining 8.0 g of the material is placed in another 20-mL portion of water, 2.0 g dissolves. On evaporation of the water, this material melts at 73.7°C. Is the original material a pure substance or a mixture?

16. Write the symbols for
   (A) iron
   (B) calcium
   (C) cobalt

17. Name
   (A) K
   (B) Pb
   (C) Ag

Elementary Atomic Structure

A = mass number

Z = atomic number

^A_Z symbol

18. (A) For the ion $_{19}^{39}K^+$, how many protons, how many electrons, and how many neutrons are present?
    (B) Which of these particles—proton, electron, neutron—has the smallest mass?

19. (A) What is the charge on a sodium ion?
    (B) What is the charge on a sodium nucleus?
    (C) What is the charge on a sodium atom?

20. (A) What is the atomic number of sodium?
    (B) How many protons are in the sodium nucleus?
    (C) How many protons are in the sodium atom?

21. An atom has a net charge of −1. It has 18 electrons and 20 neutrons. Give
    (A) its number of protons
    (B) its atomic number
    (C) its mass number
    (D) the charge on its nucleus
    (E) its isotopic symbol

22. Carbon occurs in nature as a mixture of atoms of which 98.89% have a mass of 12.0000 u and 1.11% have a mass of 13.00335 u. Calculate the atomic mass of carbon.

Ionic and Covalent Bonding

23. Write the formula of a (the) compound expected when each of the following pairs of elements is combined.
    (A) carbon and chlorine
    (B) sodium and sulfur
    (C) nitrogen and lithium

24. What is the formula of the compound corresponding to the combination of each of the following pairs?
    (A) Al and S
    (B) ClO_3^- and Co^{3+}
    (C) PO_4^{3-} and Mg^{2+}
    (D) Na and Cl_2

25. Determine the charges of the ions in parentheses.
    (A) $Ca(C_2O_4)$
    (B) $Ca(C_2H_3O_2)_2$
    (C) $Mg_3(AsO_3)_2$

26. The formula of calcium pyrophosphate is $Ca_2P_2O_7$. Determine the formulas of sodium pyrophosphate and aluminum pyrophosphate.

Electron Dot Structures and the Octet Rule

27. Write all possible octet structural formulas for
    (A) CH_4O
    (B) C_2H_3F

28. Draw electron dot diagrams for
    (A) SO_3
    (B) SO_3^{2-}
    (C) Na_2SO_3
    (D) H_2SO_3
    (E) $COCl_2$ (the central atom is C)
    (F) NH_4^+

29. Write electron dot diagrams and line structures for
    (A) phosphorus trichloride, PCl_3
    (B) carbon monoxide, CO
    (C) hydroxide ion, OH^-

30. Write electron dot diagrams for
    (A) CN^-
    (B) H_2CO
    (C) BrO_3^-

CHAPTER **3**

Periodic Table

There is a Periodic Table on pages 182–184.

Periodic Trends

31. List the symbols of the elements in each of the following groups.
- (A) alkali metals
- (B) noble gases
- (C) halogens
- (D) alkaline earth metals

32. Which is the most nonmetallic of the following?
- (A) Be
- (B) B
- (C) Al
- (D) Ga
- (E) Mg

33. Use the words at the right to complete the sentences at the left.
- (A) Bonds between nonmetal atoms are _____.   ionic
- (B) Bonds between a metal atom and a nonmetal   covalent
  atom are _____.   both types
- (C) In some compounds, _____ are formed.   neither

34. In the periodic table, note that for the main group elements, if the periodic group of the elements is even, any monatomic ion and any oxyanion of the element have charges that are even numbers, while if the periodic group number is odd, the respective charges are odd. Is this generalization true for elements that are not in a main group of the periodic table? Give examples.

35. Write the formula of a binary compound of fluorine with each main group element in the fourth period of the periodic table.

36. Which one of the following elements has chemical properties most like those of sulfur?
    (A) Cl
    (B) F
    (C) P
    (D) N
    (E) Se

37. (A) Locate in the periodic table the metals that form only one type of monatomic ion, and relate the charges of those ions to their classical periodic group numbers.
    (B) Repeat with the nonmetallic monatomic ions.

Inorganic Nomenclature

Tables 3.1, 3.2, and 3.3 are not given on examinations. If you do not already know the names and formulas of the oxyanions, the information in Table 3.1 is an efficient way to learn many of the most important ones. Table 3.2 gives you the rest. Learning the 8 more names and formulas in that table is a little easier than brute memorization since some of the rules of Table 3.1 apply. Similarly, Table 3.3 will be useful for learning the names of all common inorganic acids.

If you memorize the names of the six common oxyanions in the second from the last column of Table 3.1, you can learn the names of a total of 40 ions or acids with the following generalizations:

1. The charges on each line of Table 3.1 are all the same (odd for odd periodic groups and even for even groups) and the number of oxygen atoms varies by one for adjacent ions.
2. The prefix *per* mean "one more O atom" and the prefix *hypo* means "one fewer."
3. The ions memorized have three oxygen atoms each except for sulfate and phosphate ions, which have 4.
4. The acids names are related to ion names by the generalizations in Table 3.3.

Table 3.1 Some Common Oxyanions

VIIA	ClO^-	hypochlorite	ClO_2^-	chlorite	ClO_3^-	chlorate	ClO_4^-	perchlorate
VIIA	BrO^-	hypobromite	BrO_2^-	bromite	BrO_3^-	bromate	BrO_4^-	perbromate
VIIA	IO^-	hypoiodite	IO_2^-	iodite	IO_3^-	iodate	IO_4^-	periodate
VA			NO_2^-	nitrite	NO_3^-	nitrate		
VA	PO_2^{3-}	hypophosphite	PO_3^{3-}	phosphite	PO_4^{3-}	phosphate		
VIA			SO_3^{2-}	sulfite	SO_4^{2-}	sulfate		
IVA					CO_3^{2-}	carbonate		

Table 3.2 Other Common Polyatomic Anions

Formula	Name
OH^-	hydroxide
CN^-	cyanide
O_2^{2-}	peroxide
CrO_4^{2-}	chromate
$Cr_2O_7^{2-}$	dichromate
MnO_4^-	permanganate
$C_2H_3O_2^-$	acetate

Table 3.3 Naming Acids

The names of anions and of their corresponding acids are related as follows.

Ion Ending	Acid Prefix	Acid Suffix
-ide	hydro-	-ic
-ite		-ous
-ate		-ic

38. How are binary nonmetal-nonmetal compounds named? How are binary metal-nonmetal compounds named?

39. (A) Name $AlCl_3$, PCl_3, $CoCl_3$.
    (B) Explain why their names are so different.

40. Name
    (A) NaCl
    (B) CuBr
    (C) $MgCl_2$
    (D) SCl_2

41. How can you tell from the formula of a compound if it is an acid?

42. Name HCl, $HClO_2$, and $HClO_3$.

43. Write formulas for
    (A) barium nitrate
    (B) aluminum sulfate
    (C) iron(II) hydroxide

44. Name
   (A) Mg_3P_2
   (B) $Hg_2(NO_3)_2$

45. Name
   (A) SO_3^{2-}
   (B) SO_3
   (C) ClO^-
   (D) H_2SO_3
   (E) $HClO_4$
   (F) PBr_3
   (G) Na_2SO_3
   (H) $NaClO_2$
   (I) $Ba(ClO)_2$
   (J) HCl
   (K) $KClO_4$
   (L) $Al(ClO_2)_3$

46. Write formulas for
   (A) hydroiodic acid
   (B) hypoiodous acid
   (C) iodous acid
   (D) iodic acid
   (E) periodic acid
   (F) iodide ion
   (G) hypoiodite ion
   (H) iodite ion
   (I) iodate ion
   (J) periodate ion

47. For each of the following pairs of elements, write the formula for and name the binary compound that they form. Also state whether the bonding in each compound is ionic or covalent.
   (A) potassium and phosphorus
   (B) carbon and fluorine
   (C) hydrogen and sulfur
   (D) potassium and hydrogen
   (E) fluorine and nitrogen

48. Name

(A) AsO_4^{3-}

(B) SeO_4^{2-}

49. Write formulas for aluminum selenate and ammonium dichromate.

50. Name

(A) $NaHCO_3$

(B) $(NH_4)_2HPO_4$

(C) NaH_2PO_4

CHAPTER 4

Chemical Formulas

Abbreviations Used in This Chapter

AM	atomic mass	FM	formula mass
MM	molecular mass	$\mathcal{M}$	molar mass
u	atomic mass unit		

Percent Composition

51. Carbon dioxide is 27% carbon by mass. Express that fact using pounds, kilograms, grams, and atomic mass units.

52. Sulfur trioxide is 40 mole percent sulfur. Which of the following ratios correctly expresses that percentage?

$$\frac{40 \text{ kg S}}{100 \text{ kg SO}_3} \qquad \frac{40 \text{ mol S}}{100 \text{ mol SO}_3} \qquad \frac{40 \text{ L S}}{100 \text{ L SO}_3}$$

53. Determine the percent composition of $Mg_3(PO_4)_2$.

54. Calculate the percent by mass of oxygen in $Ca(ClO_3)_2$.

55. A strip of electrolytically pure copper with a mass of 3.178 g is strongly heated in a stream of oxygen until it is all converted to 3.978 g of a black oxide. What is the percent composition of this oxide?

56. What is the percent nitrogen (fertilizer rating) of
(A) NH_4NO_3
(B) $(NH_4)_2SO_4$

The Mole; Formula Calculations

57. ^{12}C is the standard for the atomic masses of atoms. What is the standard for the molecular masses of molecules? Explain.

58. What is the difference between molecular mass and molar mass?

59. (A) What is the formula mass of Na_2S?
    (B) What is the molecular mass of glucose, $C_6H_{12}O_6$?

60. (A) What is the mass of 4.00×10^{-3} mol of glucose, $C_6H_{12}O_6$?
    (B) How many carbon atoms are in 4.00×10^{-3} mol of glucose?

61. How many moles of $C_2H_4O_2$ contain 6.02×10^{23} atoms of hydrogen?

62. Which of the following, if either, contains fewer oxygen atoms? Fewer molecules?

$$1.0 \text{ g of } O_2 \text{ or } 1.0 \text{ g of ozone, } O_3$$

63. What is the mass in grams of one molecule of CH_3OH?

64. What mass of each of the constituent elements is contained in 1.000 mol of
    (A) CH_4
    (B) Fe_2O_3

    How many atoms of each element are contained in the same amount of compound?

65. A 5.82 g "silver" coin is dissolved in nitric acid. When sodium chloride is added to the solution, all the silver is precipitated as AgCl. The AgCl precipitate weighs 7.20 g. Determine the percentage by mass of silver in the coin.

66. A 1.5276-g sample of $CdCl_2$ was converted to metallic cadmium and chlorine by an electrolytic process. The mass of the metallic cadmium was 0.9367 g. If the atomic mass of chlorine is taken as 35.453 u, what is the atomic mass of Cd?

67. If 2.000 mol of calcium carbonate (100.1 g/mol) occupies a volume of 67.00 mL, what is its density?

Empirical Formulas

68. What is the empirical formula of a compound that contains 60.0% oxygen and 40.0% sulfur by mass?

69. A 2.500-g sample of uranium was heated in the air. The resulting oxide weighed 2.949 g. Determine the empirical formula of the oxide.

70. A sample of a pure compound contains 2.04 g of sodium, 2.65×10^{22} atoms of carbon, and 0.132 mol of oxygen atoms. Find the empirical formula.

71. An organic compound was found on analysis to contain 47.37% carbon and 10.59% hydrogen. The rest was presumed to be oxygen. What is the empirical formula of the compound?

72. A 1.367-g sample of an organic compound was burned in a stream of oxygen to yield 3.002 g CO_2 and 1.640 g H_2O. If the original compound contained only C, H, and O, what is its empirical formula?

Molecular Formulas

73. A compound has the following percent composition: C = 40.0%, H = 6.67%, O = 53.3%. Its molecular mass is 60.0 u. Derive its molecular formula.

74. Determine the empirical formula and the molecular formula of a hydrocarbon (a compound of carbon and hydrogen only) that has a molecular mass of 84 u and that contains 85.7% carbon.

75. A compound with molecular mass about 175 u consists of 40.0% carbon, 6.7% hydrogen, and 53.3% oxygen. What is its molecular formula?

CHAPTER 5

Background for the Structure of the Atom

Physical Background

76. Show that 6.02×10^{23} u = 1.00 g.

77. What did Rutherford's alpha-particle–scattering experiment prove?

78. In an oil drop experiment, the following charges (in arbitrary units) were found on a series of oil droplets: 2.30×10^{-15}, 6.90×10^{-15}, 1.38×10^{-14}, 5.75×10^{-15}, 3.45×10^{-15}, 1.96×10^{-14}. Calculate the magnitude of the charge on the electron (in the same units).

79. Although alpha particles have a larger charge, a stream of beta particles is deflected more than a stream of alpha particles in a given electric field. Explain this observation.

Light

$$E = h\nu = hc/\lambda$$

80. Which of the following relate to light as wave motion, to light as a stream of particles, or to both?
   (A) diffraction
   (B) interference
   (C) photoelectric effect
   (D) $E = mc^2$
   (E) $E = h\nu$

81. Which has the greater energy—a photon of violet light or a photon of red light?

82. When white light that has passed through sodium vapor is viewed through a spectroscope, the observed spectrum has a dark line at 589 nm. Explain this observation.

Bohr Theory

Ask your instructor if these data will be given on the exam.

$$R = 1.09678 \times 10^7 \text{ m}^{-1}$$

$$h = 6.63 \times 10^{-34} \text{ J} \cdot \text{s}$$

$$c = 3.00 \times 10^8 \text{ m/s}$$

$$\frac{1}{\lambda} = R\left(\frac{1}{n_1^2} - \frac{1}{n_2^2}\right)$$

83. Calculate the wavelengths of the first line and the series limit for the Lyman series for hydrogen.

84. If the energy difference between the excited state of an atom and its ground state is 4.4×10^{-19} J, what is the wavelength of the photon in this transition?

85. The third line in the Balmer series corresponds to an electronic transition between which Bohr orbits in hydrogen?

86. Calculate the frequency of light emitted for an electron transition from the fifth to the second orbit of the hydrogen atom. In what region of the spectrum does this light occur?

87. A photon was absorbed by a hydrogen atom in its ground state, and the electron was promoted to the fifth orbit. When the excited atom returned to its ground state, visible and other quanta were emitted. In this process, radiation of what wavelength *must* have been emitted? Explain.

CHAPTER 6

Electronic Structure of the Atom

Shells, Subshells, Orbitals

Note: The quantum numbers m_l and m_s are denoted in some texts as m and s, respectively.

88. (A) What are the possible values of l for an electron with $n = 3$?
(B) What are the possible values of m_l for an electron with $l = 2$?
(C) What are the possible values of m_s for an electron with $m_l = 0$?

89. Give the set of quantum numbers that describe an electron in a $3p$ orbital.

90. How many electrons can be placed
(A) in the shell with $n = 2$
(B) in the shell with $n = 3$
(C) in the shell with $n = 3$ before the first electron enters the shell with $n = 4$?

Electronic Structures of Atoms and Ions

91. What is the detailed electronic configuration of phosphorus?

92. Write the electronic configuration and identify the periodic group for the elements whose atomic numbers are 6, 16, 26, 36, 56.

93. Write the detailed electronic configuration of each of the following elements: Li, Na, K, Rb, Fe, F.

94. Write the electronic configurations of S^{2-} and Fe^{2+}.

95. Write the detailed electronic configurations for the following ions: Br^-, Ca^{2+}, Fe^{2+}, P^{3-}.

96. Write the detailed electronic configurations for Cl^- and Ni^{2+}. How many unpaired electrons are in the Ni^{2+} ion?

97. Write the electronic structures for
   (A) Ar and S^{2-}
   (B) Fe and Ni^{2+}
   (C) Which pair is isoelectronic? Explain.
   (D) Are *any* ions of a transition element isoelectronic with free elements?

98. Explain, using the copper atom and the copper(II) ion as examples, why the electronic configurations of some ions are rather easily predictable despite the fact that the electronic configurations of the corresponding atoms do not obey the Aufbau principle.

99. Write the electronic structure of each of the following ions:
   (A) Pb^{2+}
   (B) Tl^+
   (C) Sn^{2+}
   (D) Explain, on the basis of these ions, what is meant by the term *inert pair*. Is the pair truly inert, or merely low in reactivity?

Consequences of Electronic Structure

100. Which properties of the elements depend on the electronic configuration of the atoms and which do not?

101. Explain, in terms of electronic configuration, why the halogens have similar chemical properties. Why do they not have identical properties?

102. Account for the great chemical similarity of the lanthanide elements (atomic numbers 57 to 71).

103. (A) Select the largest species in each group: Ti^{2+}, Ti^{3+}, Ti; F^-, Ne, Na^+.
   (B) Select the species with the largest ionization potential in each group: Na, K, Rb; F, Ne, Na.
   (C) Explain why Pd and Pt have such similar sizes.

104. Which ion has the smallest radius? Li^+ Na^+ K^+ Be^{2+} Mg^{2+}

105. Which ion or atom has the largest radius? Mg Na Na^+ Mg^{2+} Al

106. The single covalent radius of P is 110 pm. What do you predict for the single covalent radius of Cl?

107. The ionization energies of Li and K are 5.4 and 4.3 eV, respectively. What do you predict for the ionization energy of Na?

108. Which of these elements is expected to have the lowest first ionization energy?   Sr As Xe S F

109. Explain why, for sodium, the second ionization energy is so much larger than the first, whereas, for magnesium, the magnitude of the difference between the first and second ionization energies is much less.

110. Explain why the first ionization energy for copper is higher than that for potassium, whereas the second ionization energies are in the reverse order.

111. All the lanthanide elements form stable compounds containing the 3+ cation. Of the few other ionic forms known, Ce forms the stablest 4+ series of ionic compounds and Eu the stablest 2+ series. Account for these unusual ionic forms in terms of their electronic configurations.

CHAPTER 7

Bonding

Bond Lengths and Bond Energies

112. Calculate the bond lengths in

    (A) NH_3

    (B) SCl_2

    (C) CH_2Cl_2

113. Arrange C—C, C=C, and C≡C in order of

    (A) increasing bond energy

    (B) increasing bond length

114. The As—Cl bond distance in $AsCl_3$ is 220 pm. Estimate the single-bond covalent radius of As.

115. Two substances having the same molecular formula, C_4H_8O, were examined in the gaseous state by electron diffraction. The carbon-oxygen distance was found to be 143 pm in compound A and 124 pm in compound B. What can you conclude about the structures of these two compounds?

Table 7.1 Covalent-Bond Radii, pm   0.1 nm = 1 Å = 100 pm

Single-Bond Radii						Multiple-Bond Radii	
H	28	P	110	Te	137	C=	67
C	77	As	121	F	64	C≡	61
Si	117	Sb	141	Cl	99	N=	63
Ge	122	O	66	Br	114	N≡	55
Sn	140	S	104	I	133		
N	70	Se	117				

116. Using data from Table 7.1, calculate the length of a molecule of C_2H_2 (a linear molecule).

117. The average C—C bond energy is 343 kJ/mol. What do you predict for the Si—Si single-bond energy?

118. In what ways does the periodic variation of the electronegativities of the elements differ from the periodic variation of ionization energies?

119. Calculate the electronegativity of nitrogen from the single-bond energies of N, 128 kJ/mol, F, 151 kJ/mol, and the energy of N—F, 225 kJ; and the electronegativity of fluorine, 4.0. Compare the result with the tabulated electronegativity of nitrogen, and suggest a reason for any difference.

Dipole Moment

120. Distinguish between a polar bond and a polar molecule. To which does the word *dipole* refer?

121. Would Br_2 or ICl be expected to have the higher boiling point?

122. Arrange in order of increasing dipole moment: BF_3, H_2S, H_2O.

123. (A) Can a molecule have a dipole moment if it has no polar covalent bonds?
     (B) How is it possible for a molecule to have polar bonds but no dipole moment?

Other Intermolecular Forces

124. Arrange the following types of interactions in order of increasing strength: covalent bond, van der Waals force, hydrogen bonding, dipole attraction.

125. List properties of water that stem from hydrogen bonding.

126. Predict the order of increasing boiling points of the noble gases.

127. Which one in each of the following pairs is expected to exhibit hydrogen bonding?
     (A) CH_3CH_2OH and CH_3OCH_3
     (B) CH_3NH_2 and CH_3SH

128. Which of the following is expected to have the highest melting point: PH_3 NH_3 $(CH_3)_3N$? Explain why.

Resonance

129. Draw two or more electron dot diagrams showing resonance in a CO_3^{2-} ion.

130. NO_2 gas is paramagnetic at room temperature. When a sample of the gas is cooled below 0°C, its molecular mass increases and it loses its paramagnetism. When it is reheated, the behavior is reversed.

    (A) Using electron dot structures, write an equation that accounts for these observations.

    (B) How does this phenomenon differ from resonance?

131. Which one(s) of the following structures *cannot* represent resonance forms for (diamagnetic) NNO?

    (A) :N̈::N::Ö:

    (B) :N:::N:Ö:

    (C) :N̈:N:::O:

    (D) :N̈::O::N̈:

    (E) :Ṅ::N::Ȯ:

Geometry of Molecules

132. Characterize the geometry of

    (A) PCl_5

    (B) H_2S

    (C) CO_2

    (D) BCl_3

    (E) H_2O

    State whether each molecule has a finite (nonzero) dipole moment.

133. PCl_5 has the shape of a trigonal bipyramid, whereas IF_5 has the shape of a square pyramid. Account for this difference.

134. Predict the geometry of

    (A) H_3O^+

    (B) $CH_2{=}NH$

    (C) ClO_2^-

    (D) NH_4^+

    (E) N_2H_4

135. Deduce the shape of
    (A) SO_3
    (B) SO_3^{2-}
    (C) BF_3
    (D) BF_4^-
    (E) NF_3

136. Which one of each of the following pairs is expected to have the larger bond angle?
    (A) H_2O and NH_3
    (B) SF_2 and BeF_2
    (C) BF_3 and BF_4^-
    (D) PH_3 and NH_3
    (E) NH_3 and NF_3

137. Two different bond lengths are observed in the PF_5 molecule, but only one bond length is observed in SF_6. Explain the difference.

CHAPTER 8

Bonding Theory

Valence Bond Theory

138. Compare the shapes of a p orbital and an sp hybrid orbital. Which one has a greater directional orientation? Explain.

139. (A) Give the geometry of dsp^2 and sp^2 hybrid orbitals.
   (B) Give the hybrid type for linear, octahedral, and trigonal bipyramidal molecules.

140. Describe the promotion and the hybrid orbitals in
   (A) each carbon atom in acetylene, $HC\equiv CH$
   (B) SF_6

141. Indicate the type of hybrid orbitals of the underlined atom and the molecular geometry.
   (A) $\underline{Be}Cl_2$
   (B) $\underline{C}Cl_4$
   (C) $\underline{C}_2F_4$
   (D) $\underline{S}F_6$
   (E) $\underline{B}Cl_3$
   (F) $H\underline{C}N$

142. Deduce the hybridization of the central atom and the geometry of each of the following.
   (A) NH_3
   (B) C_2H_4
   (C) ClO_3^-

143. The carbon-carbon double-bond energy in C_2H_4 is 615 kJ/mol, and the carbon-carbon single-bond energy in C_2H_6 is 347 kJ/mol. Why is the double-bond energy appreciably less than twice the single-bond energy?

144. The two —CH_2 groups in C_2H_4 do not rotate freely around the bond connecting them, but the two —CH_3 groups in C_2H_6 have almost an unhindered rotation around the C—C bond. Why?

145. (A) Which molecule, AX_3, AX_4, AX_5, AX_6, is most likely to have a trigonal bipyramidal structure?
(B) If the central atom, A, has no lone pairs, what type of hybridization will it have?

Molecular Orbital Theory

146. Describe the orbital configuration (boundary surface diagram) for the orbital types.
(A) $3d_{xy}$
(B) σ_{2p_x}
(C) π^*_{2px}
(D) sp_x

147. (A) Draw and label an energy level diagram for molecular orbitals formed from the $2p$ orbitals of the atoms of a diatomic molecule with 14 electrons.
(B) Use this diagram to give a molecular orbital description of the CO molecule.
(C) Briefly compare the CO molecule to the CN^- ion.

148. What is the bond order in NO?

149. Explain why NO^+ is more stable toward dissociation into its atoms than NO, whereas CO^+ is less stable than CO.

150. Has the peroxide ion, O_2^{2-}, a longer or shorter bond length than O_2? Explain.

151. Distinguish between nonbonding orbitals and antibonding orbitals.

152. Which, if either, of the following is paramagnetic?

(A) O_2^{2-}

(B) BN

153. If the internuclear axis in the diatomic molecule AB is designated as the z axis, what are the various pairs of s, p, or d atomic orbitals that can be combined to form π_x orbitals?

154. The bonding σ_{2s} orbital has a higher energy than the antibonding σ_{1s}^{*} orbital. Why is the former a bonding orbital, while the latter is antibonding?

CHAPTER 9

Organic Molecules

Organic Nomenclature and Classification

155. Name the following compounds.

(A) 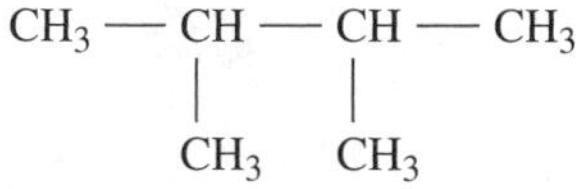
$$CH_3 — CH — CH — CH_3$$
$$\qquad\quad |\qquad\ \ |$$
$$\qquad\quad CH_3\ \ CH_3$$

(B) 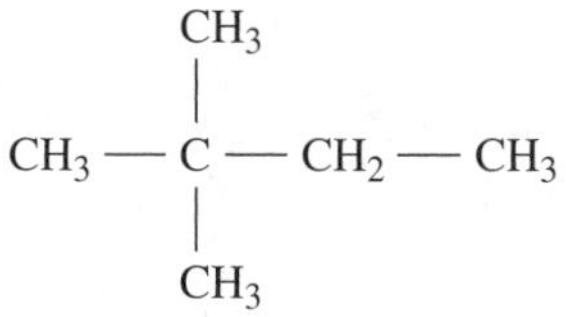
$$\qquad\quad CH_3$$
$$\qquad\quad |$$
$$CH_3 — C — CH_2 — CH_3$$
$$\qquad\quad |$$
$$\qquad\quad CH_3$$

(C)
$$\qquad\qquad\quad CH_3$$
$$\qquad\qquad\quad |$$
$$CH_3 — CH_2 — C — CH_3$$
$$\qquad\qquad\quad |$$
$$\qquad\qquad\quad CH_3$$

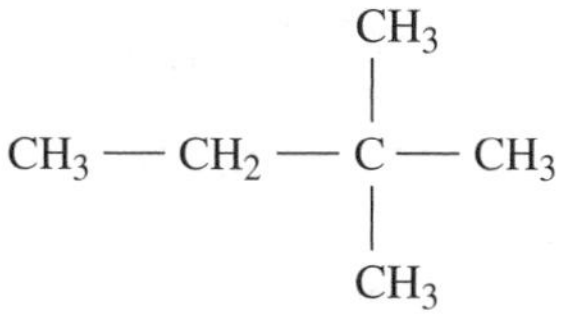

156. Write formulas for
(A) 2-methyl-3-ethylhexane
(B) 2,2-dimethylhexane
(C) 2,3,6-trimethylheptane

157. Explain why each of the following names for organic molecules is incorrect, and give the correct name.
(A) 1-methylbutane
(B) 2-methylbutane
(C) 3-methylbutane
(D) 4-methylbutane

158. What is the maximum number of other atoms to which an atom of each of the following can be bonded in organic compounds?

(A) hydrogen
(B) carbon
(C) chlorine
(D) oxygen
(E) nitrogen

159. Name: $CH_3CH_2CHCH_2CH_2CH_2CH_3$

$$CH_3 \text{---} CH \text{---} CH_3$$

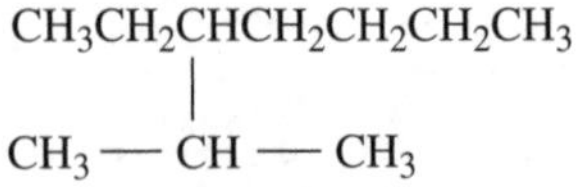

160. What class of compounds is represented by the type formula ROR′? Using CH_3 and/or C_6H_5 as the radicals, write formulas for three compounds that correspond to this type formula.

161. Name the following compounds.

(A) 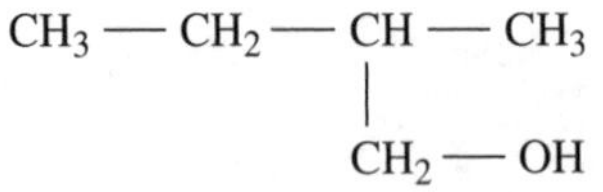
$$CH_3 \text{---} CH_2 \text{---} CH \text{---} CH_3$$
$$CH_2 \text{---} OH$$

(B) 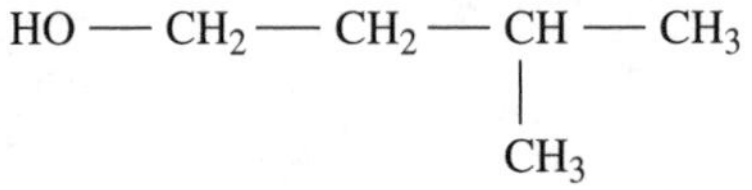
$$HO \text{---} CH_2 \text{---} CH_2 \text{---} CH \text{---} CH_3$$
$$CH_3$$

(C) 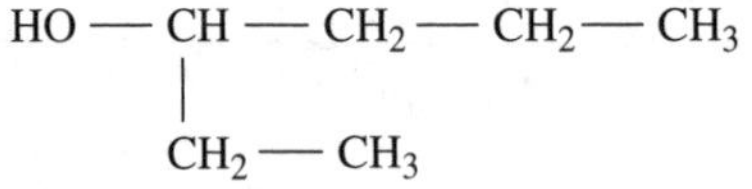
$$HO \text{---} CH \text{---} CH_2 \text{---} CH_2 \text{---} CH_3$$
$$CH_2 \text{---} CH_3$$

(D) 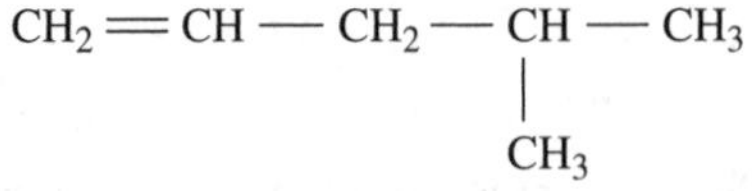
$$CH_2 \text{==} CH \text{---} CH_2 \text{---} CH \text{---} CH_3$$
$$CH_3$$

(E) $CH_3 \text{---} CH \text{==} CH \text{---} CH_2 \text{---} CH_3$

162. Name the following compounds.

(A) $CH_3CCH_2CH_2Cl$
　　　　$||$
　　　　CH_2

(B) $CH_3(CH_2)_6CO_2H$

(C) $CH_3CHCH_2CH_2CH_3$
　　　　$|$
　　　　CHO

163. (A) Explain why the name *butanol* is not specific, whereas the name *butanone* represents one specific compound.

(B) Is the name *pentanone* specific?

164. Write formulas and systematic names for

(A) ethylamine
(B) propionaldehyde
(C) butanone
(D) ethyl propionate
(E) butyl formate
(F) bromobenzene
(G) acetylene
(H) phenylacetylene

165. Write formulas for

(A) 2-butene
(B) methyl ethyl ether
(C) propanal
(D) 2-propanol

166. Write a formula for each of the following.

(A) 4-ethylheptane
(B) 4-propylheptane

167. Molecules containing which of the functional groups act in an aqueous solution?

(A) as Brønsted bases
(B) as Brønsted acids

168. For each class of simple organic compounds, write a formula for the compound in the class that contains the fewest carbon atoms, and name each compound.

169. Identify the class (Problem 168) of the following compounds, and name each.

(A) $CH_3CH_2OCH_3$
(B) $CH_3CH_2CO_2H$
(C) CH_3CONH_2
(D) $CH_3CO_2CH_3$

170. What unbranched hydrocarbon is isomeric with 2-methyl-3-ethylhexane?

Structural Isomerism

171. Write formulas for all the structural isomers of cyclobutane.

172. How many isomers correspond to the formula $C_4H_{10}O$?

173. Write formulas for all isomers of pentane, C_5H_{12}.

174. Name all nine structural isomers of $C_4H_8I_2$.

175. Write formulas for all structural isomers with the molecular formula C_5H_{10}.

Geometric Isomerism

176. Write formulas for all structural and geometric isomers of C_4H_8.

177. Which of the following compounds can exist as geometric isomers?
CH_2Cl_2, CH_2Cl—CH_2Cl, $CHBr$=$CHCl$, CH_2Cl—CH_2Br

178. Select the pair(s) of geometric isomers from the following.

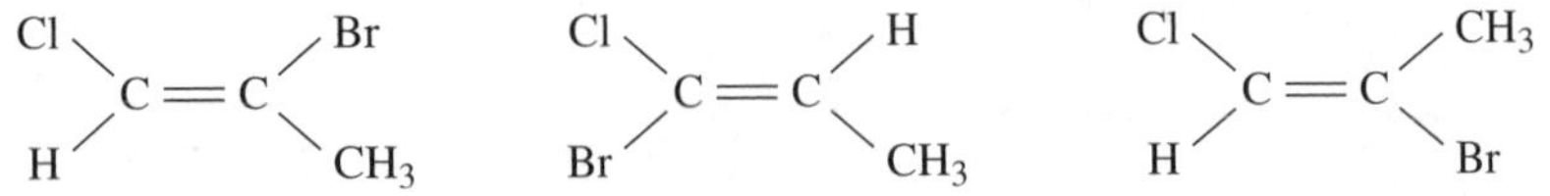

CHAPTER 10

Chemical Equations

Balancing Chemical Equations

179. In a *balanced* chemical equation, what does the absence of any coefficient imply?

180. Balance the following, using the smallest integral coefficients.

$$FeS_2 + O_2 \rightarrow Fe_2O_3 + SO_2$$

181. Write balanced chemical equations for the following reactions.

zinc sulfide + oxygen gas $\rightarrow$ zinc oxide + sulfur dioxide

nitric acid + copper(II) carbonate $\rightarrow$
water + carbon dioxide + copper(II) nitrate

182. Balance the following equations.
(A) $BCl_3 + P_4 + H_2 \rightarrow BP + HCl$
(B) $C_2H_2Cl_4 + Ca(OH)_2 \rightarrow C_2HCl_3 + CaCl_2 + H_2O$
(C) $(NH_4)_2Cr_2O_7 \rightarrow N_2 + Cr_2O_3 + H_2O$
(D) octane (C_8H_{18}) plus oxygen yields carbon monoxide plus water

183. Convert the following into balanced chemical equations.
(A) $NCl_3 + H_2O \rightarrow NH_3 + HOCl$
(B) $PCl_3 + H_2O \rightarrow H_3PO_3 + HCl$
(C) $SbCl_3 + H_2O \rightarrow Sb(O)Cl + HCl$

Prediction of Products

184. Classify simple inorganic chemical reactions into five types (leaving more complex reactions to be discussed later). Give an example of each.

185. Each of the following sets of reactants most probably represents which type of reaction as presented in Problem 184?
  (A) two elements
  (B) two compounds
  (C) one compound
  (D) one element plus one compound
  (E) an acid and a base

186. Classify each of the following substances as acid, base, acid anhydride, or basic anhydride.
  (A) H_2SO_3
  (B) NH_3
  (C) LiOH
  (D) Li_2O
  (E) Cl_2O_3
  (F) BaO
  (G) CO_2
  (H) CrO

187. Which type of simple chemical reaction (Problem 184) depends on
  (A) the relative activities of two metals or two nonmetals?
  (B) the solubilities of the reactants and/or products?

188. (A) Name five ions or classes of ions whose compounds are virtually all soluble in water.
  (B) Which metal chlorides are insoluble in water?

189. Write complete and balanced equations for each of the following reactions. If there is no reaction, write "no reaction." State the type of reaction, as described in Problem 184.
  (A) $H_2O_2 \xrightarrow{\text{heat}}$
  (B) $H_2 + O_2 \xrightarrow{\text{spark}}$
  (C) sodium plus chlorine
  (D) iron(II) chloride plus silver nitrate
  (E) calcium oxide plus carbon dioxide

190. Write complete and balanced equations for each of the following reactions. If there is no reaction, write "no reaction."

(A) $Zn + FeCl_2 \rightarrow$

(B) $F_2 + NaCl \rightarrow$

(C) $NaCl + AgNO_3 \rightarrow$

(D) $AgCl + NaNO_3 \rightarrow$

(E) $H_3PO_4 + NaOH$ (limited) $\rightarrow$

(F) $KHSO_4 + KOH \rightarrow$

(G) $SO_2 + H_2O \rightarrow$

(H) $P_2O_3 + H_2O \rightarrow$

191. Complete and balance equations for the following reactions.

(A) $AgNO_3 + Cu \rightarrow$

(B) $Cl_2 + Al \rightarrow$

(C) $NaI + Cl_2 \rightarrow$

(D) $CO + O_2 \rightarrow$

(E) $HCl + Ba(OH)_2 \rightarrow$

(F) $BaCl_2 + K_2SO_4 \rightarrow$

(G) $KClO_3 \xrightarrow{\text{heat}}$

192. Write balanced chemical equations for the following reactions.

(A) $NH_4Cl + NaOH \rightarrow$

(B) $NaC_2H_3O_2 + HCl \rightarrow$

(C) calcium and hot water

(D) $Li + O_2 \rightarrow$

(E) $Mg + HCl(aq) \rightarrow$

(F) $C_6H_6 + O_2$ (excess) $\rightarrow$

(G) $HCl(aq) + H_2SO_4(aq) \rightarrow$

Net Ionic Equations

193. Explain why net ionic equations are written only for reactions in solution, almost always aqueous solutions.

194. Explain why sodium ion is almost always a spectator ion.

195. What species that appear in complete equations are omitted from net ionic equations?

196. Write net ionic equations for the processes that occur when solutions of the following electrolytes are mixed.

(A) $AgClO_3(aq)$ and $Na_2S(aq)$

(B) $(NH_4)_3PO_4(aq)$ and $HgSO_4(aq)$

197. Write net ionic equations for each of the following complete equations.

(A) $K_2SO_4(aq) + BaCl_2(aq) \rightarrow BaSO_4(s) + 2\,KCl(aq)$

(B) $2\,HNO_3(aq) + Ca(HCO_3)_2(aq) \rightarrow Ca(NO_3)_2(aq) + 2\,H_2O + 2\,CO_2(g)$

(C) $2\,HClO_3 + Fe(OH)_2(s) \rightarrow Fe(ClO_3)_2 + 2\,H_2O$

(D) $HCl + NaHCO_3 \rightarrow NaCl + CO_2 + H_2O$

(E) $Fe_2(SO_4)_3(aq) + Fe(s) \rightarrow 3\,FeSO_4(aq)$

(F) $NaOH + NH_4Cl \rightarrow NH_3 + H_2O + NaCl$

198. Write balanced net ionic chemical equations for the following reactions.

(A) $NaHCO_3 + NaOH \rightarrow Na_2CO_3 + H_2O$

(B) $Zn + Hg_2Cl_2 \rightarrow ZnCl_2 + 2\,Hg$

(C) $2\,AgNO_3 + H_2S \rightarrow Ag_2S + 2\,HNO_3$

(D) $HClO_3 + NaOH \rightarrow NaClO_3 + H_2O$

(E) $Zn + 2\,HCl \rightarrow ZnCl_2 + H_2$

(F) $CuCl_2 + H_2S \rightarrow CuS + 2\,HCl$

CHAPTER 11

Stoichiometry

Quantities in Chemical Reactions

199. Calculate the number of moles of $Ca(HCO_3)_2$ required to prepare 2.50 mol of CO_2 according to the equation

$$Ca(HCO_3)_2 + 2\,HCl \rightarrow CaCl_2 + 2\,CO_2 + 2\,H_2O$$

200. Calculate the mass of $BaCO_3$ produced when excess CO_2 is bubbled through a solution containing 0.205 mol of $Ba(OH)_2$.

201. $C_{12}H_{22}O_{11} + 12\,O_2 \rightarrow 12\,CO_2 + 11\,H_2O$

(A) What mass of CO_2 is produced per gram of sucrose used?
(B) How many moles of oxygen gas are needed to react with 1.00 g of sucrose?

202. (A) Calculate the mass of $KClO_3$ necessary to produce 1.23 g of O_2.
(B) What mass of KCl is produced along with this quantity of oxygen?

203. $2\,NaIO_3 + 5\,NaHSO_3 \rightarrow 3\,NaHSO_4 + 2\,Na_2SO_4 + H_2O + I_2$

How much $NaIO_3$ and how much $NaHSO_3$ must be used to produce 1.00 kg of iodine?

204. Cu_2S reacts upon heating in oxygen to produce copper metal and sulfur dioxide.

(A) Write a balanced chemical equation for the reaction.
(B) What mass of copper can be obtained from 501 g of Cu_2S by this process?

205. Calculate the mass of SO_2 that can be prepared by the treatment of 105 g of Na_2SO_3 with HCl. The two other products are familiar.

206. If 4.00 g of a mixture of calcium carbonate and sand is treated with an excess of hydrochloric acid, and 0.880 g of CO_2 is produced, what is the percent of $CaCO_3$ in the original mixture?

207. A 10.20-mg sample of an organic compound containing carbon, hydrogen, and oxygen only was burned in excess oxygen, yielding 23.1 mg of CO_2 and 4.72 mg of H_2O. Calculate the empirical formula of the compound.

Limiting Quantities

208. Calculate the number of moles of each of the compounds produced and of the excess reactant when 1.50 mol of A and 0.50 mol of B are allowed to react according to the following general equation

$$3\,A + 2\,B \rightarrow C + 3\,D$$

209. Consider the reaction $2\,C + O_2 \rightarrow 2\,CO$. Calculate the number of moles of CO that will be produced by treatment of 0.179 mol C with 0.0810 mol O_2.

210. What mass of Na_2SO_4 will be formed by addition of 17.0 g of $NaHCO_3$ in aqueous solution to an aqueous solution containing 0.400 mol of H_2SO_4?

211. Calculate the number of moles of each of the products and of the excess reactant when 0.250 mol of PCl_5 and 2.00 mol of H_2O are allowed to react, yielding H_3PO_4 and HCl.

212. Calculate the number of moles of each solute in the final solution after 0.750 mol of aqueous $BaCl_2$ and 0.700 mol of aqueous $AgNO_3$ are mixed.

213. Calculate the number of moles of each ion in the final solution after 0.750 mol of aqueous $BaCl_2$ and 0.70 mol of aqueous $AgNO_3$ are mixed.

CHAPTER **12**

Measures of Concentration

Molarity

$$M = moles/liter = mmol/mL$$

214. What volume of 1.71 M NaCl solution contains 0.200 mol NaCl?

215. What volume of 3.00 M NaOH (40.0 g/mol) can be prepared with 84.0 g NaOH?

216. What is the molarity of NaOH in a solution that contains 24.0 g NaOH dissolved in 300 mL of solution?

217. Calculate the volume of 2.50 M sugar solution that contains 0.400 mol sugar.

218. What volume of water must be added to 200 mL of 0.650 M HCl to dilute the solution to 0.200 M?

219. What volume of 0.30 M Na_2SO_4 solution is required to prepare 2.0 L of a solution 0.40 M in Na^+?

220. Calculate the final concentration of HNO_3 if 0.20 mol HNO_3 is added to a beaker containing 2.0 L of 1.1 M HNO_3 and enough pure water is added to give a final volume of 3.0 L.

221. What mass of solute is required to prepare 1.000 L of 1.000 M $Pb(NO_3)_2$? What is the molarity of the solution with respect to each of the ions?

222. Exactly 100 g NaCl is dissolved in sufficient water to give 1500 cm^3 solution. What is the molarity?

223. Which two of the following solutions contain approximately equal hydrogen ion concentrations?
(A) 50 mL 0.10 M HCl + 25 mL H$_2$O
(B) 50 mL 0.10 M HCl + 50 mL H$_2$O
(C) 50 mL 0.10 M H$_2$SO$_4$ + 25 mL H$_2$O
(D) 25 mL 0.10 M H$_2$SO$_4$ + 50 mL H$_2$O

224. If 40.00 mL of 1.600 M HCl and 60.00 mL of 2.000 M NaOH are mixed, what are the molarities of Na$^+$, Cl$^-$, and OH$^-$ in the resulting solution? Assume a total volume of 100.0 mL.

225. What volume of 0.300 M H$_2$SO$_4$ is required to neutralize 200.0 mL of 0.500 M NaOH?

226. What volume of 3.000 M HCl should be added to react completely with 12.35 g NaHCO$_3$?

$$HCl + NaHCO_3 \rightarrow NaCl + CO_2 + H_2O$$

227. Zn + H$_2$SO$_4$ → ZnSO$_4$ + H$_2$. What volume of 3.00 M H$_2$SO$_4$ is required to react with 10.0 g of zinc?

228. Calculate the concentrations of all species remaining in solution after treatment of 50.0 mL of 0.300 M HCl with 50.0 mL of 0.400 M NH$_3$.

229. Calculate the concentration of an HCl solution if 2.50 mL of the solution neutralized 4.50 mL of 3.00 M NaOH.

230. A 40.0-mL sample of Na$_2$SO$_4$ solution is treated with excess BaCl$_2$. If the mass of the precipitated BaSO$_4$ is 1.756 g, what was the molarity of the Na$_2$SO$_4$ solution?

231. What mass of copper will be replaced from 2.00 L of 1.50 M CuSO$_4$ solution by 40.0 g of aluminum?

232. Calculate the mass of CuS and the concentration of H$^+$ ion produced by bubbling excess H$_2$S into 1.00 L of 0.10 M CuCl$_2$ solution. The equation is Cu^{2+} + H$_2$S(g) → CuS(s) + 2 H$^+$.

Mole Fraction

In this book,

$$x = \text{mole fraction}$$

$$n = \text{number of moles}$$

$$x(A) = \frac{\text{moles A}}{\text{total number of moles}}$$

233. Show algebraically that the sum of the mole fractions of all components of a solution must equal 1.00.

234. (A) What is the mole fraction of H_2 in a gaseous mixture containing 1.0 g H_2, 8.0 g O_2, and 16.0 g CH_4?
   (B) Determine the mole fractions of both substances in a solution containing 36.0 g water and 46.0 g glycerin, $C_3H_5(OH)_3$.

235. The density of a 2.00 M solution of acetic acid (60.0 g/mol) in water is 1.02 g/mL. Calculate the mole fraction of acetic acid.

Molality

Caution: Be sure to distinguish carefully between molarity and molality. There is only a one-letter difference in the spelling of the words, and the difference in the symbols is the difference between *M* and *m*, but:

   Molarity is the number of moles of solute per *liter* of *solution*.

   Molality is the number of moles of solute per *kilogram* of *solvent*.

236. What is the molality of a solution that contains 20.0 g cane sugar, $C_{12}H_{22}O_{11}$, dissolved in 125 g water?

237. (A) What mass of $CaCl_2$ should be added to 300 mL water to make up a 2.46 m solution?
   (B) The molality of a solution of ethyl alcohol, C_2H_5OH, in water is 1.54 mol/kg. What mass of alcohol is dissolved in 2.50 kg water?

238. A solution contains 57.5 mL ethyl alcohol (C_2H_5OH) and 600 mL benzene (C_6H_6). What mass of alcohol is in 1000 g benzene? What is the molality of the solution? Density of C_2H_5OH is 0.800 g/mL; of C_6H_6, 0.900 g/mL.

239. For a sulfuric acid solution of density 1.198 g/mL, containing 27.0% H_2SO_4 by mass, calculate the

(A) molarity
(B) molality

240. Calculate the molalities and the mole fractions of acetic acid in two solutions prepared by dissolving 120 g acetic acid

(A) in 100 g water
(B) in 100 g ethanol (C_2H_5OH)

241. Calculate the molality of an aqueous solution with mole fraction of solute

$$x = 0.0450$$

242. What is the mole fraction of the solute in a 1.00 m aqueous solution?

243. What mass of ammonium chloride is dissolved in 100 g water in each of the following solutions?

(A) 1.10 m NH_4Cl solution
(B) A solution that is 75.0% water by mass
(C) A solution with a mole fraction of 0.150 NH_4Cl

244. An aqueous solution labeled 35.0% $HClO_4$ had a density of 1.251 g/cm^3. What are the molarity and molality of the solution?

245. The density of 10.0% by mass KCl solution in water is 1.06 g/mL. Calculate the molarity, molality, and mole fraction of KCl in this solution.

CHAPTER 13

Gases

Boyle's Law, Charles' Law, Combined Gas Law

246. A sample of gas occupies 2.00 L at 760 torr. Calculate the volume it will occupy at 1.25 atm and the same temperature.

247. Calculate the volume that 4.00 L of gas at 0°C will occupy if its temperature is changed to 100°C at the same pressure.

248. Calculate the temperature at which a 2.00-L sample of gas at 27°C would occupy 3.00 L if its pressure were changed from 1.00 atm to 800 torr.

249. A sample of gas at 80°C and 785 torr occupies 350 mL. What volume will the gas occupy at STP?

Moles of Gas and the Ideal Gas Law

250. Compute the mass of a 6.00-L sample of ammonia gas, (NH_3), at STP.

251. Calculate the volume of 0.3000 mol of a gas at 60°C and 0.821 atm.

252. Calculate the volume of 8.40 g N_2 at 100°C and 800 torr.

253. What is the volume of 18.0 g of pure water at 1.00 atm and 4°C?

254. A gas cylinder contains 370 g oxygen gas at 3.00 atm pressure and 25°C. What mass of oxygen would escape if first the cylinder were heated to 75°C and then the valve were held open until the gas pressure was 1.00 atm, the temperature being maintained at 75°C?

255. At 300 K and 1.00 atm pressure, the density of gaseous HF is 3.17 g/L. Explain this observation, and support your explanation by calculations.

Dalton's Law

$$P_{total} = P_1 + P_2 + P_3 \cdots$$

256. A 10.0-L vessel containing 1.00 g H_2 at 27.0°C is connected to a 20.0-L vessel containing 88.0 g CO_2, also at 27.0°C. When the gases are completely mixed, what are the partial pressures and total pressure (in atm)?

257. A 200-mL flask contained oxygen at 220 torr, and a 300-mL flask contained nitrogen at 100 torr at the same temperature. The two flasks were then connected so that each gas filled their combined volume. Assuming no change in temperature, what was the partial pressure of each gas in the final mixture and what was the total pressure?

258. If 3.00 L oxygen gas is collected over water at 27°C when the barometric pressure is 787 torr, what is the volume of the dry gas at 0°C and 1.00 atm, assuming ideal behavior? Vapor pressure of water at 27°C is 27 torr.

259. Exactly 100 mL of oxygen is collected over water at 23°C and 800 torr. Compute the volume of the dry oxygen at STP. Vapor pressure of water at 23°C is 21.1 torr.

260. What mass of oxygen is contained in 10.5 L of oxygen measured over water at 25°C and 740 torr? Vapor pressure of water at 25°C is 24 torr.

261. Determine the molar mass of a gas if 1.80 g occupies 560 mL at 0°C and 1.00 atm.

Reactions Involving Gases

262. Calcium metal reacts with hydrochloric acid to yield hydrogen and calcium chloride. Write a balanced chemical equation for the reaction. Determine the volume of hydrogen gas at 1.00 atm pressure and 18°C produced from the reaction of 12.2 g of calcium with excess HCl.

263. What volume of oxygen, at 18°C and 750 torr, can be obtained from 110 g $KClO_3$?

264. What mass of $KClO_3$ is needed to prepare 1.80 L of oxygen collected over water at 22°C and 760 torr? Vapor pressure of water at 22°C is 19.8 torr.

265. All the oxygen in $KClO_3$ can be converted to O_2 by heating it in the presence of a catalyst.

    (A) What volume of oxygen measured at 20°C and 0.996 atm pressure can be prepared from 450 g $KClO_3$?

    (B) If the oxygen were collected over water, what would be the volume of gas collected? At 20°C the vapor pressure of water is 17.5 torr.

266. Calculate the change in pressure when 1.040 mol NO and 20.0 g O_2 in a 20.0-L vessel originally at 27°C react to produce the maximum quantity of NO_2 possible according to the equation $2NO + O_2 \rightarrow 2NO_2$.

267. What volume of CO_2 at 1.00 atm and 0°C can be produced from 112 g CO and 48.0 g O_2? What reagent is present in excess?

268. Into a 3.00-L container are placed 0.123 mol of O_2 and 0.273 mol of C at 25°C.

    (A) What is the initial pressure?

    (B) If the carbon and oxygen react as completely as possible to form CO, what will the final pressure in the container be at 25°C?

van der Waals Equation

$$P = \frac{nRT}{V - nb} - \frac{n^2 a}{V^2}$$

Table 13.1 van der Waals Constants

Gas	a L$^2 \cdot$ atm/mol^2	b L/mol
CO_2	3.59	0.0427
NH_3	4.17	0.0371
N_2	1.39	0.0391

269. Calculate the pressure of 0.60 mol NH_3 gas in a 3.00-L vessel at 25°C

    (A) with the ideal gas law

    (B) with the van der Waals equation, using data from Table 13.1

270. Using the van der Waals equation, calculate the pressure exerted by 10.0 mol carbon dioxide in a 2.00-L vessel at 47°C. Repeat the calculation using the equation of state for an ideal gas. Compare these results with the experimentally observed pressure of 82 atm.

271. For a given number of moles of gas, show that the van der Waals equation predicts greater deviation from ideal behavior

    (A) at high pressure rather than low pressure at a given temperature
    (B) at low temperature rather than high temperature at a given pressure

272. Compare the values of the van der Waals constants for NH_3 and N_2 (Table 13.1). Explain why the value of a is larger for NH_3 but that of b is larger for N_2.

Basics of Kinetic Molecular Theory

$$\overline{KE} = \tfrac{1}{2} m u^2$$

273. Which postulate(s) of the kinetic molecular theory can be used to justify

    (A) Dalton's law of partial pressures?
    (B) Graham's law of effusion?

274. What is the relationship between k, the Boltzmann constant, and R, the ideal gas law constant?

275. Identify the postulate of the kinetic molecular theory most closely identified with the van der Waals constant a, and explain the relationship. Repeat the procedure for the constant b.

276. If separate samples of argon, neon, nitrogen, and ammonia, all at the same initial temperature and pressure, are expanded to double their original volumes with no heat exchange with the surroundings, which would require the greatest quantity of heat to restore the original temperature?

277. When CO_2 under high pressure is released from a fire extinguisher, particles of solid CO_2 are formed, despite the low sublimation temperature of CO_2 at 1.0 atm pressure ($-77°C$). Explain this phenomenon.

278. Consider a sample of gas in a fixed-volume container. From the arguments of the kinetic molecular theory, show that quadrupling the absolute temperature causes a quadrupling in pressure.

279. Predict from the kinetic molecular theory what the effect will be on the pressure of a gas inside a cubic box if the length of each of its sides is reduced from l to $l/2$. Assume no change in temperature.

Graham's Law

Remember this equation.

$$\frac{r_1}{r_2} = \sqrt{\frac{M_2}{M_1}}$$

280. Calculate the ratio of rates of effusion of H_2 and O_2, both at 0°C and 1 atm pressure.

281. A large cylinder of helium filled at a certain high pressure had a small orifice through which helium escaped into an evacuated space at the rate of 6.4 mmol/h. How long would it take for 10 mmol of CO to leak through a similar orifice if the CO were confined at the same pressure?

282. A space capsule is filled with neon gas at 1.00 atm and 290 K. The gas effuses through a pinhole into outer space at such a rate that the pressure drops by 0.30 torr/s. If the capsule were filled with ammonia at the same temperature and pressure, what would be the rate of pressure drop?

CHAPTER 14

Solids and Liquids

Crystal Structure

283. Explain why an end-centered unit cell cannot be cubic. What is the highest possible symmetry for this type of unit cell?

284. Explain why uncharged atoms or molecules never crystallize in simple cubic structures.

285. Explain why a hexagonal closest-packed structure and a cubic closest-packed structure for a given element would be expected to have the same density.

286. The unit cell of the cesium chloride structure has a cesium ion at the center of a cube and chloride ions at its corners (or vice versa). To what system does cesium chloride belong? Is the unit cell compound or simple?

287. (A) Calculate the number of CsCl formula units in a unit cell. See Problem 286.
(B) What is the coordination number of each type of ion?

288. If the density of crystalline CsCl is 3.988 g/cm^3, calculate the volume effectively occupied by a single CsCl ion pair in the crystal.

289. What sort of electromagnetic radiation has wavelengths comparable to the dimensions found in crystalline solids?

290. Explain why the two allotropic forms of carbon—diamond and graphite—have such different electrical conductivities.

Crystal Energies

291. The melting point of quartz, one of the crystalline forms of SiO_2, is 1610°C, and the sublimation point of CO_2 is −79°C. How similar do you expect the crystal structures of these two substances to be?

292. Contrast the visible changes that occur while heat is added to

(A) an ice cube
(B) a bar of chocolate
(C) Which type of behavior is characteristic of an amorphous solid?
(D) List three other common amorphous materials.

293. Contrast the submicroscopic structures of crystalline and amorphous solids.

Liquids

294. (A) Distinguish between the triple point and the freezing point of a substance.
(B) For most pure substances, which is apt to be higher, the triple point or the freezing point?

295. All other factors being equal, which will cool to room temperature faster—a closed container of water at 100°C or an open container of water at 100°C? Explain your answer.

296. Explain why water would completely fill a fine capillary tube that is held vertically and is open at both ends when one end is immersed in water.

297. An astronaut in an orbiting spaceship spilled a few drops of his drink, and the liquid floated around the cabin. In what geometric shape was each drop most likely to be found? Explain.

298. (A) On the phase diagram for water, Figure 14.1, what feature represents the equilibrium of solid water and water vapor?

(B) What phase change(s) occur(s) when a sample at point E is heated at constant pressure until point F is reached?

(C) What is the temperature at which line $\overline{BC}$ intersects the 1 atm line called?

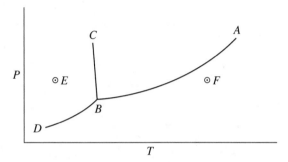

Figure 14.1

299. Knowing that the density of ice is less than that of water, explain why the slope of the solid-liquid equilibrium line in the phase diagram of water (Figure 14.1) is in accord with Le Châtelier's principle.

CHAPTER 15

Oxidation and Reduction

Oxidation Number: Oxidizing and Reducing Agents

300. What is the oxidation number of sulfur in each of the following?
  (A) S^{2-}
  (B) H_2SO_4
  (C) $S_2O_3^{2-}$
  (D) CS_2
  (E) S_8
  (F) $Na_2S_4O_6$
  (G) S_2Cl_2

301. What is the oxidation state of hydrogen in each of the following?
  (A) HCl
  (B) H^+
  (C) NaH
  (D) H_2
  (E) $LiAlH_4$

302. In the process of drying your face with a towel, what may be considered the wetting agent? The drying agent? What happens to the wetting agent? To the drying agent? In an oxidation-reduction reaction, what happens to the oxidizing agent? To the reducing agent?

303. Which one(s) of the following involve redox reactions?
  (A) Burning of gasoline
  (B) Evaporation of water
  (C) Human respiration
  (D) Preparation of metals from their ores
  (E) Production by lightning of nitrogen oxides from nitrogen and oxygen in the atmosphere
  (F) Production by electric discharge of ozone (O_3) from O_2
  (G) Reaction of H_2SO_4 with NaOH

304. Predict the highest and lowest possible oxidation states of each of the following elements:

(A) Ta
(B) Te
(C) Tc
(D) Ti
(E) Tl

305. Considering oxidation states, predict the formulas of the compounds that are likely to be produced by the combination of

(A) Cs and I
(B) S and F

306. Which of the following are examples of disproportionation reactions? What criteria determine whether a reaction is a disproportionation?

(A) $Ag(NH_3)_2^+ + 2H^+ \rightarrow Ag^+ + 2NH_4^+$
(B) $Cl_2 + 2OH^- \rightarrow ClO^- + Cl^- + H_2O$
(C) $CaCO_3 \rightarrow CaO + CO_2$
(D) $2HgO \rightarrow 2Hg + O_2$
(E) $Cu_2O + 2H^+ \rightarrow Cu + Cu^{2+} + H_2O$
(F) $CuS + O_2 \rightarrow Cu + SO_2$
(G) $2HCuCl_2 \xrightarrow{\text{dilute with } H_2O} Cu + Cu^{2+} + 4Cl^- + 2H^+$

307. In the following reaction, identify the species oxidized, the species reduced, the oxidizing agent, and the reducing agent.

$$Fe^{2+} + 2H^+ + NO_3^- \rightarrow Fe^{3+} + NO_2 + H_2O$$

308. Name the compounds in each of the following:

(A) $FeCl_2$, $FeCl_3$
(B) UCl_3, UF_6
(C) Cu_2O, CuO
(D) HNO_2, HNO_3

309. Write the formulas for sulfide ion, sulfite ion, sulfate ion, and thiosulfate ion (in which one oxygen atom of sulfate is replaced by a sulfur atom). Compare the oxidation numbers of sulfur in these species. Use this exercise to distinguish between the concepts of oxidation number and charge. Is there any direct relationship between the oxidation number of sulfur and the charge on the ion?

310. Hydrogen peroxide, H_2O_2, may act as an oxidizing agent or a reducing agent. Explain why this behavior is possible. Write an equation for the disproportionation of H_2O_2.

311. Which of the following are
(A) very good oxidizing agents
(B) good reducing agents
(C) neither

MnO_4^-, I^-, Cl^-, Ce^{4+}, $Cr_2O_7^{2-}$, Na, Na^+, CrO_4^{2-}, HNO_3, Fe^{2+}, F_2, F^-

312. Which of the following equations represent oxidation-reduction reactions? Identify each oxidizing agent and each reducing agent.
(A) $K + O_2 \rightarrow KO_2$
(B) $H_2O_2 + KOH \rightarrow KHO_2 + H_2O$
(C) $Ca(HCO_3)_2 \xrightarrow{\text{heat}} CaCO_3 + CO_2 + H_2O$
(D) $Cr_2O_7^{2-} + 2OH^- \rightarrow 2CrO_4^{2-} + H_2O$
(E) $H_2O_2 \rightarrow H_2O + \frac{1}{2}O_2$

Balancing Redox Equations

Note: Very many ways are available to balance redox equations systematically. Choose one (or two) methods; learn and use it (them) exclusively. Since various instructors and textbooks use different methods, problems are solved by two different methods here.

Ion-Electron Method

1. Identify the elements oxidized and reduced, and balance the numbers of atoms of each in a separate half-reaction equation.
2. In acid solution, balance any oxygen atoms by adding water to the other side.
3. Balance any hydrogen atoms by adding hydrogen ions to the other side.
4. Balance the charge by adding electrons to the more positive (or less negative) side.
5. Balance any other atoms by inspection, as done in earlier chapters.
6. In basic solution, do the previous steps, then add sufficient OH^- ions to neutralize any H^+ (or H_3O^+) present. Finally, subtract enough water from both sides to eliminate it on one side.

Oxidation-State Method

1. Identify the elements oxidized and reduced, balance the numbers of atoms of each, and determine the (total) change in oxidation state for each.
2. If necessary, balance the changes in oxidation state by multiplying the numbers of moles in the oxidation and/or the reduction by the lowest appropriate integer(s).
3. Balance the other species by inspection.

The oxidation-state method seems easier, but the ion-electron method has direct application in electrochemical reactions.

313. (A) If a problem does not state "acid solution" or "basic solution" explicitly, how can you tell which type of solution should be assumed?
　　(B) State whether each of the following should be balanced in acidic or basic solution.
　　　(i) $HNO_3 + Fe^{2+} \rightarrow Fe^{3+} + NO_2$
　　　(ii) $NH_3 + MnO_4^- \rightarrow MnO_2 + NO_2$
　　　(iii) $Fe^{3+} + H_2O_2 \rightarrow Fe^{2+} + O_2$
　　　(iv) $Cr(OH)_2 + I_2 \rightarrow Cr(OH)_3 + I^-$

314. Balance the following oxidation-reduction equation.

$$KMnO_4 + KCl + H_2SO_4 \rightarrow MnSO_4 + K_2SO_4 + H_2O + Cl_2$$

315. Balance the following oxidation-reduction equations by the ion-electron method.
　　(A) $K_2Cr_2O_7 + HCl \rightarrow KCl + CrCl_3 + H_2O + Cl_2$
　　(B) $FeCl_2 + H_2O_2 + HCl \rightarrow FeCl_3 + H_2O$
　　(C) $Cu + HNO_3(dil) \rightarrow Cu(NO_3)_2 + H_2O + NO$

316. Balance the following equations.
　　(A) $Na_2C_2O_4 + KMnO_4 + H_2SO_4 \rightarrow$
$$K_2SO_4 + Na_2SO_4 + H_2O + MnSO_4 + CO_2$$
　　(B) $I_2 + Na_2S_2O_3 \rightarrow Na_2S_4O_6 + NaI$
　　(C) $NH_3 + O_2 \rightarrow NO + H_2O$
　　(D) $CuO + NH_3 \xrightarrow{heat} N_2 + H_2O + Cu$

CHAPTER 16

Properties of Solutions

Raoult's Law and Vapor Pressure Lowering

Know this equation.

$$P = xP°$$

317. The vapor pressure of pure liquid solvent A is 0.80 atm. When a nonvolatile substance B is added to the solvent, its vapor pressure drops to 0.60 atm. What is the mole fraction of component B in the solution?

318. The vapor pressure of pure water at 26°C is 25.21 torr. What is the vapor pressure of a solution that contains 20.0 g glucose, $C_6H_{12}O_6$, in 70.0 g water?

319. The vapor pressure of pure water at 25°C is 23.76 torr. The vapor pressure of a solution containing 5.40 g of a nonvolatile substance in 90.0 g water is 23.32 torr. Compute the molar mass of the solute.

320. Calculate the mole fraction of toluene in the vapor phase that is in equilibrium with a solution of benzene and toluene having a mole fraction of toluene 0.500. The vapor pressure of pure benzene is 119 torr; that of toluene is 37.0 torr.

Freezing Point Lowering and Boiling Point Elevation

Know the equations; the values will be provided.

$$\Delta t_f = K_f m \quad \Delta t_b = K_b m \quad \text{For water, } K_b = 0.512°C/m \text{ and } K_f = 1.86°C/m$$

321. What is the freezing point of a 10.0% (by mass) solution of CH_3OH in water?

322. (A) When 10.6 g of a nonvolatile substance is dissolved in 740 g of ether, its boiling point is raised 0.284°C. What is the molar mass of the substance? Molal boiling point constant for ether is 2.11°C/m.

(B) An aqueous solution boils at 100.50°C. What is the freezing point of the solution?

323. When 36.0 g of a solute having the empirical formula CH_2O is dissolved in 1.20 kg of water, the solution freezes at −0.93°C. What is the molecular formula of the solute?

Osmotic Pressure

Carefully note the concentration type associated with each of the colligative properties! Use mole fraction with vapor pressure lowering, molality with freezing point depression and boiling point elevation, and molarity with osmotic pressure.

Know the following equation:

$$\pi V = nRT$$ where π is the symbol for osmotic pressure

324. Calculate the osmotic pressure of a 0.00100 M solution of a nonelectrolyte at 0°C.

325. What is the molar mass, M_A, of a solute A if the osmotic pressure of a solution containing 10.0 g/L is 10.0 torr at 27°C?

326. A 250-mL water solution containing 0.140 mol of sucrose, $C_{12}H_{22}O_{11}$, at 300 K is separated from pure water by means of a semipermeable membrane. What pressure must be applied above the solution to just prevent osmosis?

Solution of Strong Electrolytes

Know these equations:

$$\Delta t_f = iK_f m \qquad \Delta t_b = iK_b m$$

327. Explain why 0.100 m NaCl in water does *not* have a freezing point equal to

(A) −0.183°C

(B) −0.366°C

328. Of the following 0.10 m aqueous solution, which one will exhibit the largest freezing point depression? the lowest freezing point?

(A) KCl

(B) $C_6H_{12}O_6$

(C) K_2SO_4

(D) $Al_2(SO_4)_3$

(E) NaCl

329. (A) A 0.100 m solution of $NaClO_3$ freezes at $-0.3433°C$. What would you predict for the boiling point of this aqueous solution at exactly 1 atm pressure?

(B) At 0.00100 m concentration of this same salt, the electrical interferences between the ions are very small, because the ions are, on the average, far apart from each other. Predict the freezing point of this more dilute solution.

330. The freezing point of a solution composed of 10.0 g of KCl in 100 g of water is $-4.5°C$. Calculate the van't Hoff factor, i, for this solution.

CHAPTER 17

Thermodynamics

Heat, Internal Energy, Enthalpy

$$\Delta E = q + w$$

331. In a certain process, 678 J of heat is absorbed by a system while 294 J of work is done on the system. What is the change in internal energy for the process?

332. Given the following information,

$A + B \rightarrow C + D \qquad \Delta E = -10.0\ kJ$

$C + D \rightarrow E \qquad \Delta E = 15.0\ kJ$

calculate ΔE for each of the following reactions:

(A) $C + D \rightarrow A + B$
(B) $2C + 2D \rightarrow 2A + 2B$
(C) $A + B \rightarrow E$

Heat Capacity and Calorimetry

You must know this equation.

$$q = mc\Delta t$$

Table 17.1 Calorimetric Data

Specific Heat Capacity, J/g·°C		Phase Transitions for H_2O, kJ/g·°C		
Ag	0.236	ΔH_{fus}	(0°C)	0.335
Al	0.895	ΔH_{vap}	(100°C)	2.26
CO	1.04			
CO_2	0.852			
$H_2O(s)$	2.09			
$H_2O(l)$	4.184			
$H_2O(g)$	1.38			
O_2	0.922			

333. Calculate the heat necessary to raise the temperature of 40.0 g of aluminum from 20.0°C to 32.3°C.

334. A 25.0–g sample of an alloy was heated to 100.0°C and placed in a beaker containing 90.0 g of water at 25.32°C. The temperature of the water rose to a final value of 27.18°C. Neglecting heat losses to the room and the heat capacity of the beaker itself, calculate the specific heat of the alloy.

335. Assuming no loss of heat to the surroundings or to the container, calculate the final temperature when 100 g of silver at 40.0°C is immersed in 60.0 g of water at 10.0°C, using data from Table 17.1.

336. How much heat is required to raise the temperature of 10.0 g of $H_2O(s)$ from −10.0°C to $H_2O(l)$ at 10.0°C?

337. Determine the resulting temperature, t, when 150 g of ice at 0°C is mixed with 300 g of water at 50.0°C.

Law of Dulong and Petit

338. Estimate the specific heat of platinum.

339. Estimate the final temperature of a system after 1.6 mol of an unknown crystalline metal at 60.0°C was immersed in 100 g of water at 20.0°C.

340. A 40.0 g sample of a metal at 50.0°C is immersed in 100.0 g of water at 10.0°C. The final temperature of the system is 13.0°C. What is the specific heat of the metal? What is the approximate atomic mass of the metal?

Enthalpy Change

$$R = 8.314 \text{ J/mol} \cdot °C$$

$$\Delta H = \Delta E + \Delta(PV)$$

$$\sum \Delta H_{f(\text{products})} - \sum \Delta H_{f(\text{reactants})}$$

Table 17.2 Standard Enthalpies and Free Energies of Formation at 298 K

	ΔH_f°, kJ/mol	ΔG_f°, kJ/mol
$Al_2O_3(s)$	−1670	−1577
$CH_4(g)$	−94.85	−50.79
$C_2H_4(g)$	+51.9	
$CO(g)$	−110.5	−137.15
$CO_2(g)$	−393.5	−394.6
$Fe_2O_3(s)$	−822.2	
$HF(g)$	−269	
$H_2O(l)$	−285.9	−237.2
$H_2O(g)$	−241.79	−228.58
$I_2(g)$	+62.43	
$NH_3(g)$	−46.2	−16.6
$NO(g)$	+90.37	
$NO_2(g)$	+33.8	+51.9

341. The reaction of cyanamide, $NH_2CN(s)$, with oxygen was run in a bomb calorimeter, and ΔE was found to be −742.7 kJ/mol of $NH_2CN(s)$ at 298 K. Calculate ΔH_{298} for the reaction:

$$NH_2CN(s) + \tfrac{3}{2}O_2 \rightarrow N_2 + CO_2 + H_2O(l)$$

342. Using data from Table 17.2, calculate the heat released upon formation of 35.2 g of CO_2 from carbon and oxygen gas.

343. Calculate $\Delta H°$ for reduction of iron(III) oxide by aluminum (thermite reaction) at 25°C.

344. Calculate ΔH_f° of $C_6H_{12}O_6(s)$ given ΔH_{comb} of $C_6H_{12}O_6(s) = -2816$ kJ/mol

345. Calculate ΔH_f° for $C_2H_6(g)$ given:

$$2\,C_2H_6(g) + 7\,O_2(g) \rightarrow 4\,CO_2(g) + 6\,H_2O(l) \qquad \Delta H = -3119 \text{ kJ}$$

346. Calculate the enthalpy of combustion of $C_2H_4(g)$.

347. $F_2 + 2\,HCl \rightarrow 2\,HF + Cl_2 \qquad \Delta H = -353$ kJ

All substances are gaseous. ΔH_f° (HF) $= -269$ kJ/mol.

Calculate the value of ΔH_f° (HCl).

348. Which one(s) of the following equations have enthalpy changes equal to
   (A) $\Delta H_f^{\circ}(CO_2)$?
   (B) $\Delta H_{comb}(C)$?
   (C) $\Delta H_{comb}(CO)$?
   (D) $\Delta H_f^{\circ}(CO)$?
   (i) $C + O_2 \rightarrow CO_2$
   (ii) $C + \frac{1}{2}O_2 \rightarrow CO$
   (iii) $CO + \frac{1}{2}O_2 \rightarrow CO_2$

349. Two solutions, initially at 25.08°C, were mixed in an insulated bottle. One contained 400.0 mL of 0.200 M weak monoprotic acid solution. The other contained 100.0 mL of a solution having 0.800 mol NaOH/L. After mixing, the temperature rose to 26.25°C. How much heat is evolved in the neutralization of 1.00 mol of the acid? Assume that the densities of all solutions are 1.00 g/mL and that their specific heat capacities are all 4.2 J/g·K. (These assumptions are in error by several percent, but the subsequent errors in the final result partially cancel each other.)

350. The heat evolved on the complete combustion of acetylene gas, C_2H_2, at 25°C is 1299 kJ/mol. Determine the enthalpy of formation of acetylene gas.

351. What is the enthalpy of sublimation of solid iodine at 25°C?

352. Given $N_2(g) + 3\,H_2(g) \rightarrow 2\,NH_3(g) \qquad \Delta H^{\circ} = -92.4$ kJ

What is the standard enthalpy of formation of NH_3 gas?

353. The heat evolved on combustion into $CO_2(g)$ and $H_2O(l)$ of 1.000 mol C_2H_6 is 1541 kJ, and of 1.000 mol C_2H_4 is 1411 kJ. Calculate ΔH of the following reaction: $C_2H_4 + H_2(g) \rightarrow C_2H_6$.

354. Calculate the enthalpy of combustion of 100.0 g of CO at 125°C.

355. In a certain experiment, when 12.00 g of carbon reacted with a limited quantity of oxygen, 241 kJ of heat was produced. Calculate the number of moles of CO and the number of moles of CO_2 produced.

$$C + O_2 \rightarrow CO_2 \qquad \Delta H_f = -393.5 \text{ kJ}$$

$$C + \tfrac{1}{2}O_2 \rightarrow CO \qquad \Delta H_f = -110.5 \text{ kJ}$$

Free Energy Change and Entropy

You should know this information.

$$R = 8.314 \text{ J/mol} \cdot \text{K}$$

$$\Delta G = \Delta H - \Delta(TS)$$

$$= \Delta H - T\Delta S \quad \text{(constant } T\text{)}$$

$$\Delta G = \Delta G_{f\text{(products)}} - \Delta G_{f\text{(reactants)}}$$

356. Molar entropies are quoted in $J/\text{mol} \cdot K$; molar heat capacities are quoted in $J/\text{mol} \cdot K$ or $J/\text{mol} \cdot °C$. Explain.

357. (A)   For the reaction at 298 K:

$$2A + B \rightarrow C$$

$\Delta H = 100$ kJ   and $\Delta S = 0.050$ kJ/K. At what temperature will the reaction become spontaneous, assuming ΔH and ΔS to be constant over the temperature range?

(B)   For the reaction at 298 K

$$A(g) + B(g) \rightarrow E(g)$$

$$\Delta E = -3.00 \text{ kJ} \qquad \Delta S = -10.0 \text{ J/K}$$

Calculate ΔG. Predict whether the reaction may occur spontaneously, as written.

358. For the reaction $2\,Cl(g) \rightarrow Cl_2(g)$, what are the signs of ΔH and ΔS?

359. A certain reaction has a value of $\Delta H = -40\ kJ$ at 400 K. Above 400 K, the reaction is spontaneous; below that temperature, it is not. Calculate ΔG and ΔS at 400 K.

360. Using data from Table 17.2, calculate $\Delta G°$ at 298 K for the reaction

$$CH_4(g) + 2O_2(g) \rightarrow CO_2(g) + 2H_2O(l)$$

361. At its melting point, 0°C, the enthalpy of fusion of water is 6.004 kJ/mol. What is the molar entropy change for the melting of ice at 0°C?

362. Using data from Table 17.2, calculate $\Delta S_{298}°$ for the reaction of 100 g of nitrogen with oxygen according to the equation

$$N_2(g) + 2O_2(g) \rightarrow 2NO_2(g)$$

363. (A) Calculate the value of $\Delta G°$ at 25°C for the following reaction;

$$H_2(g) + CO_2(g) \rightleftharpoons H_2O(g) + CO(g)$$

(B) What is ΔG at 25°C under conditions where the partial pressures of H_2, CO_2, H_2O, and CO are 10.00, 20.00, 0.02000, and 0.01000 atm, respectively?

364. (A) The enthalpy change for a certain reaction at 298 K is -15.0 kJ/mol. The entropy change under these conditions is -7.2 J/mol·K. Calculate the free energy change for the reaction, and predict whether the reaction may occur spontaneously.

(B) For the reaction

$$2A(g) + B(g) \rightarrow 2D(g)$$

$\Delta E_{298}° = -4.50$ kJ and $\Delta S_{298}° = -10.5$ J/K. Calculate $\Delta G_{298}°$ for the reaction, and predict whether the reaction may occur spontaneously.

365. Explain in terms of thermodynamic properties why heating to a high temperature causes decomposition of $CaCO_3$ to CaO and CO_2.

CHAPTER 18

Chemical Kinetics

Rate Laws

Note: Before working the problems in this section, you should be familiar with Table 18.1 (page 154).

366. Distinguish explicitly between the rate of a given chemical reaction and the rate constant for the reaction.

367. A certain reaction between A and B is second order. Write three different rate law expressions that might possibly apply to the reaction.

368. In a certain catalytic experiment involving the Haber process, $N_2 + 3H_2 \rightarrow 2NH_3$, the rate of reaction was measured as

$$\text{Rate} = \frac{\Delta[NH_3]}{\Delta t} = 2.0 \times 10^{-4} \text{ mol} \cdot L^{-1} \cdot s^{-1}$$

If there were no side reactions, what was the rate of reaction expressed in terms of

(A) N_2?
(B) H_2?

369. What are the units of the rate constant, k, for

(A) a zero order reaction
(B) a first order reaction
(C) a second order reaction
(D) a third order reaction

370. Which of the following will produce the most product in a given time and which will react at the highest *rate*?

(A) 1 mol of A and 1 mol of B in a 1-L vessel
(B) 2 mol of A and 2 mol of B in a 2-L vessel
(C) 0.2 mol of A and 0.2 mol of B in a 0.1-L vessel

371. (A) Three 5.00–mL samples of 1.0–M reagent B were poured into three vessels containing 5.00-mL samples of A, each having the concentration tabulated below. The initial rates are also tabulated. The temperature was 25°C, and all other conditions were constant. Another set of experiments was performed in which the concentration of B was varied; these results are also tabulated. What is the order of the reaction with respect to A and with respect to B? What is the value of the rate constant?

Conc A M	Conc B M	Initial Rate M/s	Conc A M	Conc B M	Initial Rate M/s
1.0	1.0	1.2×10^{-2}	1.0	1.0	1.2×10^{-2}
2.0	1.0	2.3×10^{-2}	1.0	2.0	4.8×10^{-2}
4.0	1.0	4.9×10^{-2}	1.0	4.0	1.9×10^{-1}

(B) The reaction $A + 2B \rightarrow C + 2D$ is run three times. In the second run the initial concentration of A is double that in the first run, and the initial rate of the reaction is double that of the first run. In the third run, the initial concentration of each reactant is double the respective concentrations in the first run, and the initial rate is double that of the first run. What is the order of the reaction with respect to each reactant?

372. The hydrolysis of methyl acetate in alkaline solution,

$$CH_3COOCH_3 + OH^- \rightarrow CH_3COO^- + CH_3OH$$

followed rate = $k[CH_3OOCH_3][OH^-]$, with k equal to 0.137 $L \cdot mol^{-1} \cdot s^{-1}$ at 25°C. A reaction mixture was prepared to have initial concentrations of methyl acetate and OH^- of 0.050 M each. How long would it take for 3.0% of the methyl acetate to be hydrolyzed at 25°C?

373. The rate of the following reaction in aqueous solution, in which the initial concentration of the complex ion is 0.100 M, is to be studied

$$[Co(NH_3)_5Cl]^{2+} + H_2O \rightarrow [Co(NH_3)_5(H_2O)]^{3+} + Cl^-$$

Explain why, under these conditions, it is possible to determine the order of the reaction with respect to $[Co(NH_3)_5Cl]^{2+}$, but not with respect to water.

374. A certain reaction, $A + B \rightarrow C$, is first order with respect to each reactant, with $k = 1.0 \times 10^{-2}\ L \cdot mol^{-1} \cdot s^{-1}$. Calculate the concentration of A remaining after 100 s if the initial concentration of each reactant was 0.100 M.

375. A certain reaction, $A + B \rightarrow$ products, is first order with respect to each reactant, with $k = 5.0 \times 10^{-3}\ M^{-1} \cdot s^{-1}$. Calculate the concentration of A remaining after 100 s if the initial concentration of A was 0.100 M and that of B was 6.00 M. State any approximation made in obtaining your result.

376. For the reaction $A \rightarrow C + D$, the initial concentration of A is 0.010 M. After 100 s, the concentration of A is 0.0010 M. The rate constant of the reaction has the numerical magnitude of 9.0. What is the order of the reaction?

377. From the accompanying data for the reaction, $A \rightarrow$ products, calculate the value of k.

Time, min	[A]	Time, min	[A]
0.0	0.100	2.0	0.080
1.0	0.090	3.0	0.070

378. Substance A reacts according to a first order rate law with $k = 5.00 \times 10^{-5}\ s^{-1}$.
(A) If the initial concentration of A is 1.00 M, what is the initial rate?
(B) What is the rate after 1.00 h?

Half-Life

Note: The concept of half-life is useful for first order reactions only.

Remember this equation.

$$kt_{1/2} = \ln 2$$

379. (A) What is the half-life of a first order reaction for which $k = 7.1 \times 10^{-5}\ s^{-1}$?
(B) The half-life of a first order reaction is 2.50 h. Calculate the value of the rate constant in s^{-1}.

380. Sucrose decomposes in acid solution into glucose and fructose according to a first order rate law, with a half-life of 3.33 h at 25°C. What fraction of a sample of sucrose remains after 9.00 h?

381. Azomethane, $(CH_3)_2N_2$, decomposes with a first order rate according to the equation

$$(CH_3)_2N_2(g) \rightarrow N_2(g) + C_2H_6(g)$$

The following data were obtained for the decomposition in a 200-mL flask at 300°C

Time, min	Total Pressure, torr
0	36.2
15	42.4
30	46.5
48	53.1
75	59.3

Calculate the rate constant and the half-life for this reaction.

Collision Theory

382. In terms of reaction kinetics, explain why each of the following speeds up a chemical reaction.

(A) catalyst
(B) increase in temperature
(C) increase in concentration

383. According to Figure 18.1,

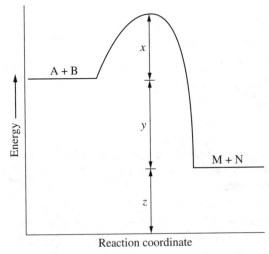

Figure 18.1

(A) The enthalpy change of the reaction $A + B \rightarrow M + N$ is represented by what value?

(B) The activation energy of the reaction $M + N \rightarrow A + B$ is represented by what value?

384. List the ways an activated complex differs from an ordinary molecule.

385. H_2 and I_2 react bimolecularly in the gas phase to form HI, and HI in turn decomposes bimolecularly into H_2 and I_2. The energies of activation for these two reactions were observed to be 163 and 184 $kJ \cdot mol^{-1}$, respectively, over the same temperature ranges near 100°C. What is ΔH for the gaseous reaction $H_2 + I_2 \rightleftharpoons 2HI$ at 100°C?

386. Distinguish between an unstable intermediate and a transition state or activated complex.

Reaction Mechanisms

387. What is the rate law for the single-step reaction $A + B \rightarrow 2C$?

388. If a rate law has the form, Rate $= k[A][B]^{1/2}$ can the reaction be an elementary process? Explain.

389. Write the equation that represents the rate law for each of the following elementary processes.

$$A + B \rightarrow C + D$$
$$2X \rightarrow E + F$$

390. What is meant by an "accepted" mechanism? Is there any such thing as a proved mechanism?

391. A possible mechanism for the reaction $2H_2 + 2NO \rightarrow N_2 + 2H_2O$ is

$$2NO \rightleftharpoons N_2O_2$$
$$N_2O_2 + H_2 \rightarrow N_2O + H_2O$$
$$N_2O + H_2 \rightarrow N_2 + H_2O$$

If the second step is rate determining, what is the rate law for this mechanism (in terms of the reactants)?

392. Nitrogen oxides and sulfur dioxide undergo the following rapid reactions in oxygen. Determine the overall reaction. What is the role of NO?

$$2\,NO + O_2 \rightarrow 2\,NO_2$$
$$NO_2 + SO_2 \rightarrow NO + SO_3$$

CHAPTER 19

Equilibrium

Le Châtelier's Principle

393. Write an "equation" for the addition of heat to a water-ice mixture at 0°C to produce more liquid water at 0°C. Which way does the equilibrium shift when you try to lower the temperature?

394. What does Le Châtelier's principle state about the effect of an addition of NH_3 on a mixture of N_2 and H_2 before it attains equilibrium?

395. What will be the effect on the following equilibrium of
    (A) the addition of more nitrogen?
    (B) an increase in temperature?
    (C) lowering the volume?

$$N_2(g) + 3H_2(g) \rightleftharpoons 2NH_3(g) + 92\text{ kJ}$$

396. (A) What is the effect of reducing the pressure by increasing the volume on the following system at 500°C?

$$H_2(g) + I_2(g) \rightleftharpoons 2HI(g)$$

(B) What effect would an increase in volume have on the following system at equilibrium at 500°C?

$$2C(s) + O_2(g) \rightleftharpoons 2CO(g)$$

Equilibrium Constants

397. Write equilibrium constant expressions for the following equations. Tell how they are related.

(A) $N_2O_4 \rightleftarrows NO_2 + NO_2$

(B) $N_2O_4 \rightleftarrows 2\,NO_2$

(C) $2\,NO_2 \rightleftarrows N_2O_4$

(D) $NO_2 \rightleftarrows \frac{1}{2}N_2O_4$

398. $X + Y \rightleftarrows Z$

A mixture of 0.120 mol of X, 0.250 mol of Y, and 0.130 mol of Z is found at equilibrium in a 1.00-L vessel.

(A) Calculate the value of K.

(B) If the same mixture had been found in a 2.00-L reaction mixture, would the value of K have been the same? Explain.

399. (A) After 0.800 mol of A and 1.60 mol of B were placed in a 1.00-L vessel and allowed to achieve equilibrium, 0.300 mol of C was found. Determine the value of the equilibrium constant.

(B) In which line of the table used to calculate the equilibrium concentrations are the values in the ratio of the coefficients of the balanced chemical equation?

400. If 1.10 mol of A and 1.40 mol of B are placed in a 1.00-L vessel and allowed to achieve equilibrium, 0.30 mol of C is found. Determine the value of the equilibrium constant.

401. (A) In a 2.00-L vessel at 488°C, H_2 and $I_2(g)$ react to form HI, with $K = 50$. If 0.100 mol of each reactant is used, how many moles of I_2 remain at equilibrium?

(B) Calculate the equilibrium CO_2 concentration if 0.0500 mol of CO and 0.0500 mol of H_2O are placed in a 1.00-L vessel and allowed to come to equilibrium at 975°C.

$$CO + H_2O(g) \rightleftarrows CO_2 + H_2 \quad K = 0.64$$

402. Calculate the number of moles of Cl_2 produced at equilibrium in a 5.00-L vessel when 0.500 mol of PCl_5 is heated to 250°C. $K = 0.0410$ mol/L.

403. If 0.20 mol of hydrogen and 1.0 mol of sulfur are heated to 90°C in a 1.0-L vessel, the following equilibrium is established:

$$H_2(g) + S(s) \rightleftarrows H_2S(g) \qquad K = 6.8 \times 10^{-2}$$

What will be the partial pressure of H_2S at equilibrium?

404. For the $N_2O_4 \rightleftarrows 2\,NO_2$ calculate the value of the equilibrium constant if initially 1.00 mol of NO_2 and 1.00 mol of N_2O_4 are placed in a 10.0-L vessel and at equilibrium 0.75 mol of N_2O_4 is left in the vessel.

405. Given the reaction $2\,XO_2 \rightleftarrows X_2O_4$ initially, 1.00 mol of XO_2 and 1.00 mol of X_2O_4 are placed in a 10.0-L vessel. At equilibrium, 0.75 mol of X_2O_4 is present in the flask. What is the value of K for the reaction?

406. A saturated solution of iodine in water is 1.30×10^{-3} M. A 0.100 M KI solution (0.100 M I^-) actually dissolves 0.0492 mol/L of iodine, most of which is converted to I_3^- because of the equilibrium $I_2(aq) + I^- \rightleftarrows I_3^-$. Assuming that the concentration of I_2 in all saturated solutions is the same, calculate the equilibrium constant for this reaction.

407. Derive the following relationship between K and K_p for a reaction in which gases are involved:

$$K_p = K(RT)^{\Delta n}$$

where Δn is the difference in the number of moles of gases between products and reactants.

Thermodynamics of Equilibrium

You must know this equation.

$$\Delta G° = -RT \ln K$$

408. If K is not numerically equal to K_p, how can both of the following equations be valid?

$$\Delta G° = -RT \ln K \qquad\qquad \Delta G° = -RT \ln K_p$$

409. Calculate ΔG and $\Delta G°$ for a reaction at equilibrium at 27°C for which $K = 10$.

410. (A)   What is the value of the equilibrium constant for the following reaction at 298 K?

$$A(g) + B(g) \rightleftarrows D(g) + C(g)$$

$$\Delta H° = -29.8 \text{ kJ} \quad \text{and} \quad \Delta S° = -0.100 \text{ kJ/K}$$

(B)   Calculate $\Delta G°$ at 490°C, for the following equilibrium:

$$H_2(g) + I_2(g) \rightleftarrows 2HI(g) + \text{heat}$$

At equilibrium: $[H_2] = 8.62 \times 10^{-4} \text{ mol/L}$

$[I_2] = 2.63 \times 10^{-3} \text{ mol/L}$

$[HI] = 1.02 \times 10^{-2} \text{ mol/L}$

CHAPTER 20

Acids and Bases

Acid-Base Theory

Note: H^+ is often used as an abbreviation for H_3O^+ in aqueous solutions. In equations where this abbreviation is used, less water is included. Thus, the following equations both represent the same reaction in aqueous solution.

$$H_2O + HCl \rightleftharpoons Cl^- + H_3O^+$$
$$HCl \rightleftharpoons Cl^- + H^+$$

Note also that not every reagent that is added to a solution will react in a manner such as given in Chapter 11. Some reagents might merely affect the position of an equilibrium.

411. (A) According to Le Châtelier's principle, what effect does sodium acetate have on an aqueous solution of acetic acid?

(B) Arrange the following 0.10 M solutions in order of increasing pH.

(i) NH_4Cl

(ii) $NaOH$

(iii) $HC_2H_3O_2$

(iv) $NaCl$

(v) $NH_3 + NH_4Cl$

(vi) NH_3

(vii) HCl

412. Write complete net ionic equations for the following processes. Which combinations of reactants will react less than 2% of the theoretically possible extent? Which one(s) will react until more than 98% of the limiting quantity is used up?

(A) $HC_2H_3O_2 + H_2O$

(B) $C_2H_3O_2^- + H_2O$

(C) $C_2H_3O_2^- + H_3O^+$

413. Write the formula for the conjugate base of each of the following acids.

(A) HCN

(B) HCO_3^-

(C) $N_2H_5^+$

414. (A) Write an equation showing the dissociation of HX, and label all acids and bases.

(B) In 1.00 L of 0.100 M solution, 0.00134 mol of ammonia dissociates. Calculate the value of the equilibrium constant.

415. In neutral solution, the amino acid glycine exists predominantly in the form $^+NH_3CH_2COO^-$. Write formulas for

(A) the conjugate base of glycine

(B) the conjugate acid of glycine

Ionization Constants

$$HC_2H_3O_2 \quad K_a = 1.8 \times 10^{-5}$$
$$NH_3 \quad K_b = 1.8 \times 10^{-5}$$

Note that when we give a concentration of a solute we mean the concentration *before* the equilibrium reaction. Otherwise we use the term "equilibrium concentration."

416. When 0.100 mol of ammonia, NH_3, is dissolved in sufficient water to make 1.00 L of solution, the solution is found to have a hydroxide ion concentration of 1.34×10^{-3} M. Calculate K_b for ammonia.

417. Calculate the hydronium ion concentration of a solution containing 0.200 mol of $HC_2H_3O_2$ in 1.00 L of solution. $K_a = 1.80 \times 10^{-5}$.

418. Calculate $[OH^-]$ of a 0.10 M NH_3 solution.

419. What (total) concentration of acetic acid is needed to give a $[H^+]$ of 3.5×10^{-3} M?

Ionization of Water

$$pH = -\log [H_3O^+] \quad \text{(Notice that this is a common log.)}$$

$$K_w = 1.0 \times 10^{-14} \quad \text{(Remember this value.)}$$

420. The number 10.92 has how many significant figures? The number 0.92 has how many? If these were pH values, how many significant figures should be reported for the corresponding hydronium ion concentrations?

421. (A) What is the pH of 10^{-2} M KOH?
(B) Calculate the pH of a solution that has a hydronium ion concentration of 6.0×10^{-8} M.

422. (A) Calculate the hydroxide ion concentration of a solution that has a pH of 11.73.
(B) Calculate the hydronium ion concentration of a 0.100 M NaOH solution.

423. Calculate the hydronium ion concentration of each of the following solutions from the given pH value.
(A) 0.00
(B) 13.85

424. Calculate the pH of 1.0×10^{-3} M solutions of each of the following.
(A) $Ba(OH)_2$
(B) NaCl

425. (A) Calculate K_a for an acid HA whose 0.10 M solution has a pH of 4.500.
(B) Calculate K_b for a base B whose 0.10 M solution has a pH of 10.500.

426. Find the pH of the solution resulting when 50 mL of 0.20 M HCl is mixed with 50 mL of 0.20 M $HC_2H_3O_2$.

427. (A) What is the pH of a solution containing 0.010 M HCl?
     (B) Calculate the change in pH if 0.020 mol $NaC_2H_3O_2$ is added to 1.0 L of this solution.

Buffer Solutions

428. Explain why a solution containing a strong base and its salt does not act as a buffer solution.

429. Which of the following combinations of solutes would result in the formation of a buffer solution?
     (A) $NaC_2H_3O_2 + HC_2H_3O_2$
     (B) $NH_4Cl + NH_3$
     (C) $HCl + NaCl$
     (D) $HCl + HC_2H_3O_2$
     (E) $NaOH + HCl$
     (F) $NaOH + HC_2H_3O_2$ in a 1:1 mole ratio
     (G) $NH_3 + HCl$ in a 2:1 mole ratio
     (H) $HC_2H_3O_2 + NaOH$ in a 2:1 mole ratio

430. Calculate the pH of a solution of 0.10 M HA and 0.20 M NaA.
     $K_a = 1.0 \times 10^{-7}$

431. It is desired to prepare a buffer solution consisting of 0.10 M $HC_2H_3O_2$ and 0.10 M $NaC_2H_3O_2$. Assuming no volume change upon the addition of the pure compounds, state what reagents and in what quantities should be added to 1.00 L of each of the following solutions to prepare the desired buffer solution.
     (A) 0.10 M $HC_2H_3O_2$
     (B) 0.20 M $HC_2H_3O_2$
     (C) 0.20 M $NaC_2H_3O_2$
     (D) 0.10 M $NaC_2H_3O_2$
     (E) 0.10 M NaOH

432. Calculate the acetic acid to acetate ion concentration ratio in a buffer solution whose pH is 7.00. Explain how it is possible to have *any* acid in a neutral solution.

433. Determine [OH⁻] of a 0.050 M solution of ammonia to which has been added sufficient NH_4Cl to make [NH_4^+] equal to 0.100.

434. Find the value of $[H_3O^+]$ in 1.0 L of solution in which are dissolved 0.080 mol $HC_2H_3O_2$ and 0.100 mol $NaC_2H_3O_2$.

435. Calculate the pH of a solution prepared by mixing 50.0 mL of 0.200 M $HC_2H_3O_2$ and 50.0 mL of 0.100 M NaOH.

436. (A) Calculate the pH of a solution of 0.10 M acetic acid.
    (B) Calculate the pH after 50.0 mL of this solution is treated with 25.0 mL of 0.10 M NaOH.

437. Determine the pH of a solution after 0.100 mol of NaOH is added to 1.00 L of a solution containing 0.150 M $HC_2H_3O_2$ and 0.200 M $NaC_2H_3O_2$. Assume no change in volume.

438. Assuming that the final volume in each case remains the same, compare the effect of adding 0.0100 mol of solid sodium hydroxide to
    (A) 1.00 L of a solution 1.8×10^{-5} M in HCl
    (B) 1.00 L of a solution containing 0.100 mol of $NaC_2H_3O_2$ and 0.100 mol of $HC_2H_3O_2$

CHAPTER 21

Heterogeneous and Other Equilibria

Solubility Product Equilibria

Table 21.1 Selected K_{sp} Values

AgCl	1.8×10^{-10}
$Mg(OH)_2$	7.1×10^{-12}
$Fe(OH)_3$	1.6×10^{-39}
$PbCl_2$	1.8×10^{-5}

439. When a sample of solid AgCl is shaken with water at 25°C, a solution containing 1.3×10^{-5} M silver ions is produced. Calculate the value of K_{sp}.

440. Calculate the solubility of $Mg(OH)_2$ in water. $K_{sp} = 7.1 \times 10^{-12}$ (Table 21.1)

441. (A) How many moles of CuI ($K_{sp} = 5 \times 10^{-12}$) will dissolve in 1.0 L of 0.010 M NaI solution?

(B) Calculate the solubility of A_2X_3 in pure water, assuming that neither kind of ion reacts significantly with water. For A_2X_3, $K_{sp} = 1.1 \times 10^{-23}$.

442. Calculate the solubility of AgCl in 0.20 M $AgNO_3$ solution.

443. Which has a greater molarity in water, AgCl or $Mg(OH)_2$? Can relative solubilities be predicted on the basis of the relative magnitudes of the K_{sp} values alone? Explain.

444. What is the solubility (in moles per liter) of $Fe(OH)_3$ in a solution of pH = 8.0? $[K_{sp} = 1.6 \times 10^{-39}]$

445. The $[Ag^+]$ of a solution is 4.0×10^{-3}. Calculate the $[Cl^-]$ that must be exceeded before AgCl can precipitate.

446. Determine whether a precipitate will form when 100 mL of 0.100 M Pb^{2+} solution is mixed with 100 mL of 0.300 M Cl^- solution.

$$K_{sp} = 1.6 \times 10^{-5}$$

447. (A) What is the maximum pH of a solution 0.10 M in Mg^{2+} from which $Mg(OH)_2$ will not precipitate?
(B) Calculate the hydroxide ion concentration of a solution after 100 mL of 0.100 M $MgCl_2$ is added to 100 mL of 0.200 M NaOH.

Competitive Reactions

448. Calculate the simultaneous solubility of CaF_2 $(K_{sp} = 3.9 \times 10^{-11})$ and SrF_2 $(K_{sp} = 2.9 \times 10^{-9})$.

449. Write equations showing all of the equilibrium reactions occurring in aqueous solutions containing each of the following sets of reagents.
(A) NaCl
(B) NaOH
(C) $NaC_2H_3O_2 + HC_2H_3O_2$
(D) $Na_2S + CuS$
(E) $NH_4Cl + NH_3 + Mg(OH)_2(s)$

450. Calculate the silver ion concentration in a solution prepared by shaking solid Ag_2S with saturated H_2S (0.10 M) in 0.15 M H_3O^+ until equilibrium is established.

$$K_{sp} = 6.3 \times 10^{-50} \quad K_1 = 9.5 \times 10^{-8} \quad K_2 = 1 \times 10^{-19}$$

451. The number of moles of CoS that dissolves per liter of solution exceeds the solubility predicted by the value of K_{sp} for CoS. Explain by means of appropriate chemical equations why this behavior is observed.

452. Equal volumes of 0.0200 M $AgNO_3$ and 0.0200 M HCN were mixed. Calculate [Ag^+] at equilibrium.

$$K_{sp} = 2.2 \times 10^{-16} \qquad K_a = 6.2 \times 10^{-10}$$

453. H_2S is bubbled into a solution containing 0.15 M Cu^{2+} until no further change takes place. Calculate the concentrations of H_3O^+ produced and of Cu^{2+} remaining in solution.

$$K_{sp} = 8.5 \times 10^{-36} \qquad K_1 = 9.5 \times 10^{-8} \qquad K_2 = 1 \times 10^{-19}$$

CHAPTER 22

Electrochemistry

Electrical Units

454. Identify each of the following symbols as it applies to electrochemistry.

V, A, J, W, I, ε, $\mathcal{F}$

455. Express each of the following combinations of electrical units as a single unit.
- (A) volt · ampere
- (B) ampere · second
- (C) volt/ampere
- (D) joule/volt
- (E) watt/ampere · ohm
- (F) joule/second
- (G) joule/ampere · second
- (H) joule/ampere^2 · second

456. How many kilojoules of energy is expended when a current of 1.00 A passes for 100 s under a potential of 115 V?

457. From the definition of a faraday, calculate the charge on one electron.

Electrolysis

458. How many faradays of electricity are required to electrolyze 1 mol $CuCl_2$ to copper metal and chlorine gas?

459. A solution of copper(II) sulfate is electrolyzed between copper electrodes by a current of 10.0 A for 1.00 h. What changes occur at the electrodes and in the solution? How many moles of copper is involved?

460. (A) Calculate the mass of mercury produced by the reduction of $Hg(NO_3)_2$ by the passage of 19,300 C of charge.
(B) If $Hg_2(NO_3)_2$ were electrolyzed under the same conditions as in part (A), what mass of mercury would be produced?

461. What is the average current (in amperes) if 100 g nickel is deposited from a nickel(II) ion solution in 3 h 20 min?

462. Determine a value for Avogadro's number, using the charge on the electron, 1.60×10^{-19} C, and the fact that 96,500 C deposits 107.9 g silver from its solution.

463. (A) A current of 2.00 A passing for 5.00 h through a molten tin salt deposits 22.2 g tin. What is the oxidation state of the tin in the salt?
(B) A current of 0.200 A is passed for 600 s through 50.00 mL of 0.1000 M NaCl. If only chlorine gas is produced at the anode and if water is reduced to hydrogen gas at the cathode, what will the hydroxide ion concentration in the solution be after the electrolysis?

Galvanic Cells

464. (A) Write equations for the half-cell reactions for the following cell from its electrochemical notation.

$$Zn \mid Zn^{2+}(1\ M) \parallel Fe^{2+}\ (1\ M),\ Fe^{3+}(1\ M) \mid Pt$$

(B) Write the equation for the cell reaction and calculate the value of ε°_{cell}.

465. Is 1.0 M H^+ solution under hydrogen gas at 1.0 atm capable of oxidizing silver metal in the presence of 1.0 M silver ion?

466. What reaction, if any, would zinc(II) ion undergo in the copper(II) half of a Daniell cell? What reaction, if any, would copper(II) ion undergo in the zinc half of a Daniell cell? Which of these ions might actually get into the other half-cell during discharge of the cell? during recharge? Explain why a Daniell cell cannot be fully recharged.

467. (A) Is tin(II) stable toward disproportionation (oxidizing and reducing itself) in noncomplexing media?
(B) Explain why aluminum metal cannot be produced by electrolysis of aqueous solutions of aluminum salts. Explain why aluminum is produced by the electrolysis of a molten mixture of Al_2O_3 and Na_3AlF_6 rather than by electrolysis of molten Al_2O_3 alone.
(C) Why are Co^{3+} salts unstable in water?

468. Neglecting electrode polarization effects, predict the principal product at each electrode in the continued electrolysis at 25°C of each of the following.

(A) 1 M $Fe_2(SO_4)_3$ with inert electrodes in 0.1 M H_2SO_4
(B) 1 M LiCl with silver electrodes
(C) molten NaF

469. What fraction of a mole of iron metal will be produced by passage of 4.00 A of current through 1.00 L of 0.100 M Fe^{3+} solution for 1.00 h? Assume that only iron is reduced. (Be careful.)

Nernst Equation

You should know this equation.

$$\varepsilon = \varepsilon° - \frac{RT}{n\mathcal{F}} \ln Q$$

470. What is the potential of an electrode consisting of zinc metal in a solution in which the zinc ion concentration is 0.0100 M at 25 °C?

471. Write the net ionic equation for the half-reaction in which HNO_3 is reduced to NO. Under which of the following sets of conditions does the half-cell potential equal the standard half-cell potential?

(A) 1 M NO_3^-, 1 M H^+, 1 atm NO
(B) 1 M NO_3^-, 4 M H^+, 1 atm NO
(C) 1 M NO_3^-, 1 M H^+, 1 atm air
(D) 1 M NO_3^-, 4 M H^+, 1 atm air

472. (A) Using the Nernst equation for the cell reaction

$$Pb + Sn^{2+} \rightarrow Pb^{2+} + Sn$$

calculate the ratio of cation concentrations for which $\varepsilon = 0$.

(B) Distinguish clearly between the meaning of $\varepsilon = 0$ and $\varepsilon° = 0$. Give an example of a reaction in which $\varepsilon° = 0$.

473. Calculate the reduction potential of a half-cell consisting of a platinum electrode immersed in 2.0 M Fe^{2+} and 0.020 M Fe^{3+} solution.

474. Write the half-cell reactions and calculate ε and $\varepsilon°$ for the following cell reaction.

$$Hg_2Cl_2(s) \rightarrow 2\,Hg(l) + Cl_2 \ (0.80 \text{ atm})$$

475. (A) Given the concentration cell Zn | Zn^{2+} (1.0 M) || Zn^{2+} (0.15 M) | Zn write equations for each half-reaction. Calculate ε. As the cell discharges, does the difference in the concentrations of the two solutions become smaller or larger?

   (B) The reversible reduction potential of pure water is –0.414 V under 1.00 atm H_2 pressure. If the reduction is considered to be

$$2\,H^+ + 2\,e^- \rightarrow H_2$$

   calculate the hydrogen ion concentration of pure water.

476. At what potential should a solution containing 1 M $CuSO_4$, 1 M $NiSO_4$, and 2 M H_2SO_4 be electrolyzed so as to deposit essentially none of the nickel and leave 1.0×10^{-9} M Cu^{2+}?

Practical Applications

477. Explain why blocks of magnesium are often strapped to the steel hulls of ocean-going ships.

478. Explain why the lead storage cell

   (A) has a relatively constant potential
   (B) has its state of charge signaled by its electrolyte density
   (C) needs no salt bridge
   (D) can be recharged

Electrochemical Equilibrium and Thermodynamics

Table 22.1 Standard Reduction Potentials at 25°C

	ε° (V)		ε° (V)
$F_2 + 2\,e^- \rightarrow 2\,F^-$	2.87	$Pb^{2+} + 2\,e^- \rightarrow Pb$	–0.126
$Co^{3+} + e^- \rightarrow Co^{2+}$	1.82	$Sn^{2+} + 2\,e^- \rightarrow Sn$	–0.14
$Cl_2 + 2\,e^- \rightarrow 2\,Cl^-$	1.36	$Ni^{2+} + 2\,e^- \rightarrow Ni$	–0.25
$O_2 + 4\,H^+ + 4\,e^- \rightarrow 2\,H_2O$	1.229	$PbSO_4 + 2\,e^- \rightarrow Pb + SO_4^{2-}$	–0.31
$Ag^+ + e^- \rightarrow Ag$	0.799	$Fe^{2+} + 2\,e^- \rightarrow Fe$	–0.44
$Fe^{3+} + e^- \rightarrow Fe^{2+}$	0.771	$Zn^{2+} + 2\,e^- \rightarrow Zn$	–0.7628
$Cu^{2+} + 2\,e^- \rightarrow Cu$	0.34	$2\,H_2O + 2\,e^- \rightarrow H_2 + 2\,OH^-$	–0.828
$Hg_2Cl_2 + 2\,e^- \rightarrow 2\,Hg + 2\,Cl^-$	0.270	$Al^{3+} + 3\,e^- \rightarrow Al$	–1.66
$AgCl + e^- \rightarrow Ag + Cl^-$	0.222	$Mg^{2+} + 2\,e^- \rightarrow Mg$	–2.37
$Sn^{4+} + 2\,e^- \rightarrow Sn^{2+}$	0.13	$Li^+ + e^- \rightarrow Li$	–3.03
$2\,H^+ + 2\,e^- \rightarrow H_2$	0.0000		

479. In an electrochemical cell, what value does the cell potential have when the reaction is at equilibrium? What is the value at equilibrium of the ratio of concentrations that is part of the Nernst equation?

480. (A) Calculate the standard free energy change for the reaction
$Zn + Cu^{2+} \rightarrow Cu + Zn^{2+}$.

 (B) Using data from Table 22.1, calculate the free energy change per mole of copper(II) ion formed in a cell consisting of a copper/copper(II) ion half-cell suitably connected to a silver/silver ion half-cell of sufficient size that the concentration of the ions is not changed from 1.00 M.

481. Write the Nernst equation in terms of free energy change instead of potential.

CHAPTER 23

Nuclear and Radiochemistry

Nuclear Particles and Nuclear Reactions

The designations for small particles are as follows.

p	proton	$_{1}^{1}\text{H}$		γ	gamma ray (photon)	$_{0}^{0}\gamma$
d	deuteron	$_{1}^{2}\text{H}$		n	neutron	$_{0}^{1}n$
α	alpha particle	$_{2}^{4}\text{He}$		β^{+}	positron	$_{+1}^{0}\beta$
β^{-}	beta particle (electron)	$_{-1}^{0}\beta$				

482. Select from the following list of nuclides.

(A) the isotopes
(B) the isobars

$$_{18}^{40}\text{Ar}, \ _{19}^{41}\text{K}, \ _{21}^{40}\text{Sc}, \ _{21}^{42}\text{Sc}, \ _{40}^{90}\text{Zr}$$

483. Show that a mass of 1.00 u is equivalent to 932 MeV of energy.

484. Do the designations $_{1}^{1}\text{H}$ and $_{0}^{1}n$ for the proton and neutron imply that these two particles are of equal mass (1 u)?

485. What are the numbers of protons, electrons, and neutrons in an atom of ^{239}Pu?

486. Write the complete nuclear symbols for natural fluorine and natural arsenic. Each has only one naturally occurring isotope.

487. Complete the following nuclear equations.

(A) $^{14}_{7}N + ^{4}_{2}He \rightarrow ^{17}_{8}O + \cdots$

(B) $^{9}_{4}Be + ^{4}_{2}He \rightarrow ^{12}_{6}C + \cdots$

(C) $^{9}_{4}Be + ^{1}_{1}p \rightarrow ^{4}_{2}\alpha + \cdots$

(D) $^{30}_{15}P \rightarrow ^{30}_{16}S + \cdots$

(E) $^{3}_{1}H \rightarrow ^{3}_{2}He + \cdots$

(F) $^{43}_{20}Ca + ^{4}_{2}\alpha \rightarrow \cdots + ^{46}_{21}Sc$

488. An alkaline earth element is radioactive. It and its two daughter elements decay by emitting alpha particles in succession. In what periodic group should the resulting element be found?

489. Which of the following nuclides is the terminal member of the naturally occurring radioactive series that begins with $^{232}_{90}Th$?

(A) $^{209}_{83}Bi$

(B) $^{208}_{82}Pb$

(C) $^{206}_{82}Pb$

(D) $^{207}_{82}Pb$

(E) $^{210}_{83}Bi$

490. If an atom of ^{235}U, after absorption of a slow neutron, undergoes fission to form an atom of ^{139}Xe and an atom of ^{94}Sr, what other type particle is produced, and how many of them?

491. Which one(s) of the following processes—alpha emission, beta emission, positron emission, electron capture—cause

(A) an increase in atomic number?

(B) a decrease in atomic number?

(C) emission of an x-ray in every case?

Half-Life

Table 23.1 Typical Half-Lives

Nuclide	Half-Life	Radiation*
$^{238}_{92}U$	4.50×10^9 y	alpha
$^{237}_{93}Np$	2.2×10^6 y	alpha
$^{14}_{6}C$	5730 y	beta
$^{90}_{38}Sr$	19.9 y	beta
$^{3}_{1}H$	12.3 y	beta
$^{140}_{56}Ba$	12.5 d	beta
$^{131}_{53}I$	8.0 d	beta
$^{140}_{57}La$	40 h	beta
$^{15}_{8}O$	118 s	beta
$^{94}_{36}Kr$	1.4 s	beta

* In most of these decay processes, gamma radiations are also emitted.

492. The half-life of $^{90}_{38}Sr$ is 20 y. If a sample of this nuclide has an initial activity of 8000 dis/min today, what will be its activity after 80 y?

493. Prove that $A/A_0 = N/N_0 = m/m_0$, where A and A_0 are activities, N and N_0 are numbers of atoms, and m and m_0 are masses, all of the same decaying nuclide.

494. ^{18}F is found to undergo 90% radioactive decay in 366 min. What is its half-life?

495. Given 2.00 kg of $^{238}_{92}U$ (half-life 4.50×10^9 y). The ultimate decay product in this series is $^{206}_{82}Pb$.
(A) What mass of $^{238}_{92}U$ is left after 3.00×10^9 y?
(B) What mass of $^{206}_{82}Pb$ is produced in this time?

496. The half-life of ^{14}C is 5730 y (Table 23.1).
(A) What fraction of its original ^{14}C would a sample of $CaCO_3$ have after 11,460 y of storage in a locality where additional radioactivity could not be produced?
(B) What fraction of the original ^{14}C would still remain after 13,000 y?

Binding Energy

Table 23.2

Particle	Symbol	Mass, u	Charge, e
Alpha	$\alpha, {}^4_2He$	4.001506	+2
Proton	p	1.0072765	+1
Neutron	n	1.0086650	0
Electron	$e^-, \beta^-, {}_{-1}\beta$	0.0005486	−1
Positron	$e^+, \beta^+, {}_{+1}\beta$	0.0005486	+1

497. When a ${}^{238}U$ nucleus disintegrates spontaneously, it forms a ${}^{234}Th$ nucleus and an α particle (${}^4He$ nucleus). Explain why the change in rest mass during this process can be calculated by subtracting the mass of a ${}^{234}Th$ *atom* plus the mass of a ${}^4H$ *atom* from the mass of the ${}^{238}U$ *atom*.

498. How much energy would have to be added to ${}^4_2He$ to make two protons and two neutrons? Use data from Table 23.2.

Radiochemistry

499. To 50.00 mL of a solution containing an unknown concentration of zinc ion was added 0.100 μCi of ${}^{62}Zn^{2+}$ in 10 mL solution, and the total volume was diluted to 100 mL with water. Precipitation of a zinc salt yielded 0.2000 g zinc in the solid phase with an activity of 0.0823 μCi. What was the original concentration of the zinc ion?

500. When radioactive sulfur is added to alkaline sodium sulfite solution, radioactive thiosulfate ion is formed. Upon adding Ba^{2+}, a precipitate of BaS_2O_3 is obtained. The precipitate is filtered and dried and is then treated with acid, producing solid sulfur, SO_2 gas, and water. The SO_2 is *not* radioactive at all. Write equations for the sequence of reactions, and comment on the structure of the thiosulfate ion as elucidated by this experiment.

Chapter 1: Measurement

1. 1 cm. (If you got a different magnitude, review exponential numbers and precedence rules on your calculator.)

2. **(A)** $(4.23 \text{ m})\left(\dfrac{1 \text{ km}}{10^3 \text{m}}\right) = 4.23 \times 10^{-3} \text{ km}$

$(4.23 \text{ m})\left(\dfrac{100 \text{ cm}}{\text{m}}\right) = 423 \text{ cm}$

$(4.23 \text{ m})\left(\dfrac{10^3 \text{ mm}}{\text{m}}\right) = 4.23 \times 10^3 \text{ mm}$

(B) $(29.66 \text{ mm})\left(\dfrac{1 \text{ cm}}{10 \text{ mm}}\right) = 2.966 \text{ cm}$

$(29.66 \text{ mm})\left(\dfrac{1 \text{ m}}{10^3 \text{mm}}\right) = 2.966 \times 10^{-2} \text{ m}$

3. **(A)** $1 \text{ m}^3 = (1 \text{ m})^3 = (100 \text{ cm})^3 = (10^2 \text{ cm})^3 = 10^6 \text{ cm}^3$

(B) $1 \text{ m}^3 = (10 \text{ dm})^3 = 10^3 \text{ dm}^3 \times 1 \text{ L/dm}^3 = 10^3 \text{ L}$

(C) $1 \text{ L} = 1 \text{ dm}^3 = (10 \text{ cm})^3 = 10^3 \text{ cm}^3$

4. Convert to decimeters, since $1 \text{ L} = 1 \text{ dm}^3$.

Volume $= 0.50 \text{ m} \times 20 \text{ cm} \times 25 \text{ mm} = 5.0 \text{ dm} \times 2.0 \text{ dm} \times 0.25 \text{ dm} = 2.5 \text{ dm}^3 = 2.5 \text{ L}$

5. **(A)** $(3.00 \times 10^2 \text{g})\left(\dfrac{10^3 \text{ mg}}{\text{g}}\right) = 3.00 \times 10^5 \text{ mg}$

(B) $(1.90 \times 10^{-2} \text{cm}^3)\left(\dfrac{1 \text{ L}}{10^3 \text{ cm}^3}\right) = 1.90 \times 10^{-5} \text{ L}$

(C) $6.21 \times 10^6 \text{ mm}$ **(D)** 3.33 cm^3 **(E)** 9.70 kg **(F)** 2.22 mL **(G)** 0.00655 g/mL
(H) 4.18 g/mL

6. **(A)** $(3.00 \times 10^2 \text{ cm}) + (0.75 \times 10^2 \text{ cm}) = 3.75 \times 10^2 \text{ cm} = 3.75 \text{ m}$

(B) $(20.0 \text{ cm})(10.0 \text{ cm})(30.0 \text{ cm}) = 6.00 \times 10^3 \text{ cm}^3 = 6.00 \text{ L}$

7. **(A)** two **(B)** three **(C)** four **(D)** two **(E)** three **(F)** an indeterminant number
(G) an infinite number

8. The answers are **(A)** 90.4 cm and **(B)** 196 g rather than **(A)** 90.42 cm and **(B)** 195.8 g. **(A)** The 2 in the hundredths column of the sum is farther to the right than the 3 of 17.3, and so it cannot be significant. It is dropped because it is less than 5.   **(B)** The 8 unrounded is not significant for the same reason, but it is over 5, so the answer is rounded up to the next higher integer.

9. (A) 142   Three significant figures may be retained, since each factor has three. **(B)** 15   Only two significant figures are retained in the answer.

10. (A) 1.066   **(B)** 750   **(C)** 0.050   **(D)** 0.6070   **(E)** 40.0

11. (A) $(4.50 \times 10^2 \text{ m}) + (3.00 \times 10^3 \text{ m}) = 3.45 \times 10^3 \text{ m}$ **(B)** $9.00 \times 10^8 \text{ cm}^2$ **(C)** $(4.50 \times 10^2 \text{ mL}) - (0.225 \times 10^2 \text{ mL}) = 4.28 \times 10^2 \text{ mL}$

12. (A) There are three significant digits in each concentration.

   (B) The values obtained on a calculator are -1.701146924, -8.701146924, and -11.70114692, respectively.

   (C) The first three digits *after the decimal point* (701) are the significant digits in each calculator value. The integer digits tell only to which power of 10 the base is raised. The values should be reported as -1.701, -8.701, and -11.701, respectively.

13. (A) $1.7 \times 10^{14} \text{ cm}$ **(B)** $2.0 \times 10^{-1} \text{ cm}$ **(C)** $10 \times 10^{13} \text{ cm}^3 = 1.0 \times 10^{14} \text{ cm}^3$ **(D)** $-2.3 \times 10^{16} \text{ cm}$

Chapter 2: Structure of Matter

Elements, Compounds, Mixtures

14. Since different parts of the combination have different properties, it must be a *mixture*.

15. It is a pure substance, since the 8.0-g portion has the same properties as the whole sample.

16. (A) Fe   **(B)** Ca   **(C)** Co

17. (A) potassium   **(B)** lead   **(C)** silver

18. (A) 19 protons, 18 electrons, 20 neutrons   **(B)** electron

19. (A) 1+   **(B)** 11+   **(C)** 0   Note that it is extremely important to read each question carefully, since a small difference in wording can make it an entirely different question.

20. The answer to each question is 11. (These seemingly different questions are all really the same question in different forms.)

21. (A) 17   **(B)** 17   **(C)** 37   **(D)** 17+   **(E)** $^{37}_{17}\text{Cl}^-$

22. $\text{AM} = (12.0000 \text{ u} \times 0.9889) + (13.00335 \text{ u} \times 0.0111) = 12.011 \text{ u}$

23. (A) CCl_4 **(B)** Na_2S **(C)** Li_3N

24. **(A)** Al_2S_3 **(B)** $Co(ClO_3)_3$ **(C)** $Mg_3(PO_4)_2$ **(D)** NaCl

25. **(A)** 2–, to balance the 2+ charge on the Ca^{2+} ion.

  (B) 1–, the two 1– ions balance the charge on the Ca^{2+} ion.

  (C) 3–, the two 3– charges balance the three 2+ charges on the Mg^{2+} ions.

26. The charge of the pyrophosphate ion must be 4– to balance the charge of two Ca^{2+} ions. We can then write $Na_4P_2O_7$ and $Al_4(P_2O_7)_3$.

27. **(A)** H : C : O : H also written as H — C — O — H **(B)**

28. **(A)** **(B)** **(C)** 2 Na⁺ **(D)**

(E) **(F)** H : N : H

29. **(A)** :Cl: P :Cl: :Cl — P — Cl: **(B)** :C :::O: :C ≡ O:

(C) [:O:H]⁻   :O — H⁻

  In **(A)** each atom has an octet, formed by sharing one pair of electrons with each neighboring atom in single bonds. In **(B)** the sharing of three pairs of electrons constitutes a triple bond. In **(C)** the hydroxide ion has a negative charge, indicating one electron in excess of those provided by the hydrogen and oxygen atoms. The hydrogen atom does not have an octet, but by sharing a pair of electrons it attains a configuration similar to that of helium.

30. **(A)** :C :::N:⁻   **(B)** H : C : H   **(C)**

Chapter 3: Periodic Table

31. **(A)** Li, Na, K, Rb, Cs, Fr   **(C)** F, Cl, Br, I, At
  (B) He, Ne, Ar, Kr, Xe, Rn   **(D)** Be, Mg, Ca, Sr, Ba, Ra

32. B (boron). It is tied for farthest right and highest of the elements listed.

33. **(A)** covalent **(B)** ionic **(C)** both types

34. No. (Fe^{2+} and Fe^{3+} both exist, for example, despite the fact that iron is a member of an even-numbered group.) Exceptions include Fe^{3+}, Mn^{2+}, Cr^{3+}, and Cu^{2+}, among many others.

35. KF; CaF_2; GaF_3; GeF_4; AsF_3 or AsF_5; SeF_2, SeF_4 or SeF_6; BrF, BrF_3, BrF_5 or BrF_7; KrF_2, KrF_4 or KrF_6

36. (E) Se, which is in the same periodic group

37. (A) Group IA, IIA, and IIIB metals, as well as Al, Zn, Cd, and Ag, form monatomic ions with charges equal to their group numbers.   **(B)** Group VIIA, VIA, and VA nonmetals form monatomic ions with charges equal to their classical group numbers minus 8. Also, hydrogen forms the hydride ion, H^-, as well as the hydrogen ion, H^+. (Instructors most often give generalities like these, but ask for specific examples on exams.)

38. Nonmetal-nonmetal compounds are usually named by giving the name of the nonmetal farthest to the left and/or farthest down in the periodic table, then giving the name of the other element with its ending changed to *ide*. The number of atoms of the second element is denoted by a prefix—mono, di, tri, tetra, penta, hexa, and so forth. Metal-nonmetal compounds are named by giving the name of the metal first. Only the element name is used for metals of groups IA and IIA, and Zn, Al, Cd, and Ag. For other metals, the oxidation state (equal to the charge on the monatomic ion) is given in parentheses attached to the name. Then the name of the nonmetal is given, with its ending changed to *ide*.

39. (A) Aluminum chloride, phosphorus trichloride, cobalt(III) chloride.   **(B)** The compounds are named using different systems. Aluminum is a metal that never varies from +3 oxidation state in its compounds, so "aluminum" chloride implies aluminum with a charge of 3+. Phosphorus is a nonmetal, and nonmetal-nonmetal compounds are named using the prefixes (tri in this case). Cobalt is a metal that forms two different ions; we must be specific about which cobalt chloride we are referring to in this compound by affixing a roman numeral to denote the charge. An older terminology uses cobaltic chloride (as opposed to cobaltous chloride) to denote the cobalt ion with the higher charge.

40. (A) sodium chloride   **(B)** copper(I) bromide   **(C)** magnesium chloride   **(D)** sulfur dichloride (a nonmetal-nonmetal compound)

41. The formulas of acids have hydrogen written first. Thus, HCl is an acid and NH_3 is not. H_2O is a notable exception. (As one gets more familiar with chemical compounds, this practice is deviated from in part; thus, organic chemists write acids in accordance with their structures, such as CH_3COOH.)

42. Hydrochloric acid, chlorous acid, and chloric acid (see Table 3.3).

43. (A) Since two negative charges are required to balance the charge on one barium ion, Ba^{2+}, and each NO_3^- ion has only one negative charge, the formula must be written $Ba(NO_3)_2$. **(B)** To achieve equal numbers of positive and negative charges, the formula must be $Al_2(SO_4)_3$. **(C)** Since iron(II) is specified, it takes two OH^- ions to supply the appropriate number of negative charges—hence, the formula is $Fe(OH)_2$.

44. (A) The ion of phosphorus must be an anion, since magnesium forms only a cation. Application of the usual method for naming binary compounds gives the name magnesium phosphide.   **(B)** Since the total charge on two NO_3^- (nitrate ions) is 2−, the total cationic charge must be 2+. Thus, the average charge per Hg is 1+, and the name is mercury(I) nitrate (or mercurous nitrate).

45. **(A)** sulfite ion

  (B) sulfur trioxide

  (C) hypochlorite ion

  (D) sulfurous acid

  (E) perchloric acid

  (F) phosphorus tribromide

  (G) sodium sulfite

  (H) sodium chlorite

  (I) barium hypochlorite

  (J) hydrogen chloride (when pure) or hydrochloric acid

  (K) potassium perchlorate

  (L) aluminum chlorite

46. **(A)** HI **(B)** HIO **(C)** HIO_2 **(D)** HIO_3 **(E)** HIO_4 **(F)** I^- **(G)** IO^- **(H)** IO_2^- **(I)** IO_3^- **(J)** IO_4^-

47. **(A)** K_3P, potassium phosphide, ionic **(B)** CF_4, carbon tetrafluoride, covalent **(C)** H_2S, hydrogen sulfide, covalent **(D)** KH, potassium hydride, ionic **(E)** NF_3, nitrogen trifluoride, covalent. Note that the element farther to the right in the periodic table is named last. Nonmetal-nonmetal compounds (except for hydrogen compounds) use the prefixes from Question 38.

48. **(A)** arsenate ion (analogous to phosphate) **(B)** selenate ion (analogous to sulfate)

49. **(A)** $Al_2(SeO_4)_3$ **(B)** $(NH_4)_2Cr_2O_7$

50. **(A)** sodium hydrogen carbonate **(B)** ammonium monohydrogen phosphate **(C)** sodium dihydrogen phosphate

Chapter 4: Chemical Formulas

51. $\dfrac{27 \text{ lb C}}{100 \text{ lb } CO_2}$ $\dfrac{27 \text{ kg C}}{100 \text{ kg } CO_2}$ $\dfrac{27 \text{ g C}}{100 \text{ g } CO_2}$ $\dfrac{27 \text{ u C}}{100 \text{ u } CO_2}$

52. Only the second ratio.

53. In 1 mol of $Mg_3(PO_4)_2$ there are

$$3 \times 24.3 = 72.9 \text{ g Mg} \qquad \frac{72.9 \text{ g Mg}}{262.9 \text{ g total}} \times 100\% = 27.7\% \text{ Mg}$$

$$2 \times 31.0 = 62.0 \text{ g P} \qquad \frac{62.0 \text{ g P}}{262.9 \text{ g total}} \times 100\% = 23.6\% \text{ P}$$

$$8 \times 16.00 = 128.0 \text{ g O} \qquad \frac{128.0 \text{ g O}}{262.9 \text{ g total}} \times 100\% = 48.7\% \text{ O}$$

$$\text{Total} = 262.9 \text{ g} \qquad\qquad \text{Total} = 100.0\%$$

You may check your answers by adding all the percentages. If the total is 100% within 0.1%, your answer *may* be correct. If the total is not 100%, you have made an error somewhere. (If the answer is 100%, you still *may* have made a mistake. Try using the wrong atomic mass for one of the elements, for example.)

54. $\quad$ Ca $\quad$ 1 × 40.0 u = $\quad$ 40.0 u

$\quad$ 2 Cl $\quad$ 2 × 35.5 u = $\quad$ 71.0 u

$\quad$ 6 O $\quad$ 6 × 16.0 u = $\underline{\quad 96.0 \text{ u}}$ $\qquad$ $\dfrac{96.0\,\text{u}}{207.0\,\text{u}} \times 100\% = 46.4\% \text{ O}$

$\qquad\qquad$ Total = 207.0 u

55. Total mass of black oxide = 3.978 g

$\quad$ Mass of copper in oxide = −3.178 g

$\quad$ Mass of oxygen in oxide = $\quad$ 0.800 g

$\quad$ Fraction of copper = $\dfrac{\text{mass of copper in oxide}}{\text{total mass of oxide}} = \dfrac{3.178 \text{ g}}{3.978 \text{ g}} = 0.7989 = 79.89\%$

$\quad$ Fraction of oxygen = $\dfrac{\text{mass of oxygen in oxide}}{\text{total mass of oxide}} = \dfrac{0.800 \text{ g}}{3.978 \text{ g}} = 0.201 = \underline{\quad 20.1\%}$

$\qquad\qquad\qquad\qquad\qquad\qquad\qquad\qquad\qquad$ Check: 100.0%

56. **(A)** $\quad$ 2 N $\quad$ 2(14.0 u) = 28.0 u $\qquad\qquad$ **(B)** $\quad$ 2 N $\quad\quad$ 28.0 u

$\qquad$ 4 H $\quad$ 4(1.008 u) = $\quad$ 4.0 u $\qquad\qquad\qquad$ 8 H $\qquad$ 8.0 u

$\qquad$ 3 O $\quad$ 3(16.0 u) = $\underline{48.0 \text{ u}}$ $\qquad\qquad\qquad\qquad$ S $\qquad$ 32.0 u

$\qquad\qquad\qquad$ Total = 80.0 u $\qquad\qquad\qquad\qquad$ 4 O $\qquad$ $\underline{64.0 \text{ u}}$

$\qquad\qquad\qquad\qquad\qquad\qquad\qquad\qquad\qquad\qquad$ Total $\quad$ 132.0 u

$\quad$ %N = $\dfrac{28.0 \text{ u}}{80.0 \text{ u}} \times 100\% = 35.0\% \text{ N}$ $\qquad$ %N = $\dfrac{28.0 \text{ u}}{132.0 \text{ u}} \times 100\% = 21.2\% \text{ N}$

57. ^{12}C. The same standard is used for all atomic, molecular, and formula masses.

58. Molecular mass is the number of atomic mass units per molecule. Molar mass is the number of grams per mole (of a molecular or ionic substance). The numerical value is the same for both, but the units are different. Thus, NH_3 has a molecular mass of 17 u and a molar mass of 17 g/mol.

59. Na_2S does not exist as discrete molecules represented by the empirical formula, but glucose does. The term *formula mass* may be used in every case to describe the relative mass of the indicated formula unit; further chemical experience is needed to define the applicability of the term *molecular mass*. Some books use the two terms interchangeably.

$\quad$ **(A)** 2 Na = 2(22.990 u) = 45.980 u $\qquad$ **(B)** 6 C = 6(12.011 u) = $\quad$ 72.066 u

$\qquad\qquad$ S $\qquad\qquad\qquad$ = 32.066 u $\qquad\qquad$ 12 H = 12(1.008 u) = $\quad$ 12.096 u

$\qquad\qquad$ Formula mass = 78.046 u $\qquad\qquad\qquad$ 6 O = 6(15.994 u) = $\underline{\quad 93.994 \text{ u}}$

$\qquad\qquad\qquad\qquad\qquad\qquad\qquad\qquad\qquad$ Molecular mass = 178.156 u

$\quad$ Note that the atomic masses are not all known to the same number of decimal places. In **(A)** there is no point in writing the atomic mass of Na as 22.98977 u, since the value for S is known to only three decimal places.

60. (A) $(4.00 \times 10^{-3}\,\text{mol})\left(\dfrac{180\,\text{g}}{\text{mol}}\right) = 0.720\,\text{g}$

(B) $(4.00 \times 10^{-3}\,\text{mol}\,C_6H_{12}O_6)\left(\dfrac{6\,\text{mol C}}{\text{mol}\,C_6H_{12}O_6}\right)\left(\dfrac{6.02 \times 10^{23}\,\text{C atoms}}{\text{mol C}}\right)$

$$= 1.44 \times 10^{22}\,\text{C atoms}$$

61. $6.02 \times 10^{23}\,\text{H atoms}\left(\dfrac{1.00\,\text{mol H}}{6.02 \times 10^{23}\,\text{H atoms}}\right)\left(\dfrac{1\,\text{mol}\,C_2H_4O_2}{4\,\text{mol H}}\right) = 0.250\,\text{mol}\,C_2H_4O_2$

62. Both have the same number of atoms. The 1.0-g sample of O_3 has fewer molecules.

63. One mole of CH_3OH contains $12.0\,\text{g C} + 4.0\,\text{g H} + 16.0\,\text{g O} = 32.0\,\text{g}\,CH_3OH$

$$\dfrac{32.0\,\text{g}}{1\,\text{mol}\,CH_3OH}\left(\dfrac{1\,\text{mol}\,CH_3OH}{6.02 \times 10^{23}\,\text{molecules}}\right) = 5.32 \times 10^{-23}\,\text{g/molecule}$$

64. (A) 1 mol C = 12.01 g C $\qquad\qquad$ 6.02×10^{23} atoms

$\qquad$ 4 mol H = 4.032 g H $\qquad\qquad$ $4(6.02 \times 10^{23}) = 2.41 \times 10^{24}$ atoms

(B) 2 mol Fe = 111.70 g Fe $\qquad\qquad$ $2(6.02 \times 10^{23}) = 1.20 \times 10^{24}$ atoms

$\qquad$ 3 mol O = 48.00 g O $\qquad\qquad$ $3(6.02 \times 10^{23}) = 1.81 \times 10^{24}$ atoms

65. $7.20\,\text{g AgCl}\left(\dfrac{1\,\text{mol AgCl}}{143.3\,\text{g AgCl}}\right)\left(\dfrac{1\,\text{mol Ag}}{1\,\text{mol AgCl}}\right)\left(\dfrac{107.9\,\text{g Ag}}{1\,\text{mol Ag}}\right) = 5.42\,\text{g Ag}$

Fraction of Ag in coin $= \dfrac{5.42\,\text{g}}{5.82\,\text{g}} \times 100\% = 93.1\%\,\text{Ag}$

66. $\qquad$ Mass of $CdCl_2 = 1.5276\,\text{g}$

$\qquad$ Mass of Cd in $CdCl_2 = 0.9367\,\text{g}$

$\qquad$ Mass of Cl in $CdCl_2 = 0.5909\,\text{g}$

$$0.5909\,\text{g Cl}\left(\dfrac{1\,\text{mol Cl}}{35.453\,\text{g Cl}}\right) = 0.01667\,\text{mol Cl}$$

$$0.01667\,\text{mol Cl}\left(\dfrac{1\,\text{mol Cd}}{2\,\text{mol Cl}}\right) = 0.008333\,\text{mol Cd}$$

$$\text{from the formula}$$

Atomic mass of Cd $= \dfrac{0.9367\,\text{g}}{0.008333\,\text{mol}} = 112.4\,\text{g/mol}$

$$= 112.4\,\text{u/atom}$$

67. $(2.000\,\text{mol})\left(\dfrac{100.1\,\text{g}}{\text{mol}}\right) = 200.2\,\text{g}$ $\qquad\qquad$ $d = \dfrac{200.2\,\text{g}}{67.00\,\text{mL}} = 2.988\,\text{g/mL}$

68. Since no specific mass of sample has been given, you may choose any total mass. A 100-g sample is a convenient choice, because then the mass of each of the elements is numerically equal to the percentage of that element in the compound.

$$60.0 \text{ g O}\left(\frac{1 \text{ mol O}}{16.0 \text{ gO}}\right) = 3.75 \text{ mol O} \quad 40.0 \text{ g S}\left(\frac{1 \text{ mol S}}{32.0 \text{ g S}}\right) = 1.25 \text{ mol S}$$

Since formulas are "always" expressed in terms of whole numbers of atoms, the mole ratio must be expressed as a ratio of integers. Therefore to obtain an integral ratio, divide the numbers of moles of the elements present by the number present in smallest amount.

$$\frac{3.75 \text{ mol O}}{1.25 \text{ mol S}} = \frac{3.00 \text{ mol O}}{1.00 \text{ mol S}}$$

The ratio of 3 mol of oxygen atoms to 1 mol of sulfur atoms corresponds to the formula SO_3.

69. 2.949 g of the oxide contains 2.500 g U and 0.449 g O.

$$2.500 \text{ g U}\left(\frac{1 \text{ mol U}}{238.03 \text{g U}}\right) = 0.01050 \text{ mol U}$$

$$0.449 \text{ g O}\left(\frac{1 \text{ mol O}}{16.00 \text{ g O}}\right) = 0.0281 \text{ mol O}$$

$$\frac{0.0281 \text{ mol O}}{0.0105 \text{ mol U}} = \frac{2.68 \text{ mol O}}{1.00 \text{ mol U}} = \frac{3(2.68 \text{ mol O})}{3(1.00 \text{ mol U})} = \frac{8.03 \text{ mol O}}{3.00 \text{ mol U}}$$

The empirical formula is U_3O_8.

Emphasis must be placed on the importance of carrying out the computations to as many significant figures as the analytical precision requires. If numbers in the ratio 2.68:1 had been multiplied by 2 to give 5.36:2 and these numbers had been rounded off to 5:2, the wrong formula would have been obtained. This would have been unjustified because it would have assumed an error of 36 parts in 450 in the analysis of oxygen. The mass of oxygen, 0.449 g, indicates a possible error of only a few parts in 450. When the multiplying factor 3 was used, the rounding off was from 8.03 to 8.00, the assumption being made that the analysis of oxygen may have been in error by 3 parts in 800; this degree of error is more reasonable.

70. $(2.04 \text{ g Na})\left(\dfrac{1 \text{ mol Na}}{23.0 \text{ g Na}}\right) = 0.0887 \text{ mol Na}$

$(2.65 \times 10^{22} \text{ C atoms})\left(\dfrac{1 \text{ mol C}}{6.02 \times 10^{23} \text{ C atoms}}\right) = 0.0440 \text{ mol C}$

and we are given 0.132 mol O. Dividing each number of moles by 0.0440, the smallest, yields the integral ratio 2:1:3; the empirical formula is Na_2CO_3.

71. $\%O = 100.00\% - 47.37\% - 10.59\% = 42.04\%$

$$(47.37 \text{ g C})\left(\frac{1 \text{ mol C}}{12.01 \text{ g C}}\right) = 3.944 \text{ mol C} \qquad (10.59 \text{ g H})\left(\frac{1 \text{ mol H}}{1.008 \text{ g H}}\right) = 10.51 \text{ mol H}$$

$$(42.04 \text{ g O})\left(\frac{1 \text{ mol O}}{16.00 \text{ g O}}\right) = 2.628 \text{ mol O}$$

$$\frac{3.944 \text{ mol C}}{2.628} = 1.501 \text{ mol C} \qquad\qquad \frac{10.51 \text{ mol H}}{2.628} = 4.000 \text{ mol H}$$

$$\frac{2.628 \text{ mol O}}{2.628} = 1.000 \text{ mol O}$$

The mole ratio is 3:8:2; the empirical formula is $C_3H_8O_2$.

72. It is necessary to find how much C and H are present in the products and, thus, in the original sample.

$$3.002 \text{ g CO}_2\left(\frac{1 \text{ mol CO}_2}{44.01 \text{ g CO}_2}\right)\left(\frac{1 \text{ mol C}}{1 \text{ mol CO}_2}\right)\left(\frac{12.01 \text{ g C}}{1 \text{ mol C}}\right) = 0.8192 \text{ g C}$$

$$1.640 \text{ g H}_2O\left(\frac{1 \text{ mol H}_2O}{18.02 \text{ g H}_2O}\right)\left(\frac{2 \text{ mol H}}{1 \text{ mol H}_2O}\right)\left(\frac{1.008 \text{ g H}}{1 \text{ mol H}}\right) = 0.1835 \text{ g H}$$

The mass of oxygen in the original sample cannot be obtained from the masses of combustion products, since the CO_2 and H_2O contain oxygen that came partly from the combined oxygen in the compound and partly from the oxygen stream used in the combustion process. The oxygen content of the sample can be obtained, however, by difference.

$$(1.367 \text{ g compound}) - (0.8192 \text{ g C}) - (0.1835 \text{ g H}) = 0.3643 \text{ g O}$$

The question can now be solved by the usual procedures. The numbers of moles of the elements in 1.367 g of compound are found to be as follows: C, 0.0682; H, 0.1820; O, 0.0228. These numbers are in the ratio 3:8:1, and the empirical formula is C_3H_8O.

73. $(40.0 \text{ g C})\left(\dfrac{1 \text{ mol C}}{12.0 \text{ g C}}\right) = 3.33 \text{ mol C} \qquad (6.67 \text{ g H})\left(\dfrac{1 \text{ mol H}}{1.008 \text{ g H}}\right) = 6.62 \text{ mol H}$

$(53.3 \text{ g O})\left(\dfrac{1 \text{ mol O}}{16.0 \text{ g O}}\right) = 3.33 \text{ mol O}$

The empirical formula is CH_2O. The mass of one empirical formula unit is

$$12.0 \text{ u} + 2(1.0 \text{ u}) + 16.0 \text{ u} = 30.0 \text{ u}$$

$$\frac{60.0 \text{ u/molecule}}{30.0 \text{ u/empirical formula unit}} = 2 \text{ empirical formula units/molecule}$$

The molecular formula is $(CH_2O)_2$ or $C_2H_4O_2$.

74. $(85.7 \text{ g C})\left(\dfrac{1 \text{ mol C}}{12.0 \text{ g C}}\right) = 7.14 \text{ mol C}$ $(14.3 \text{ g H})\left(\dfrac{1 \text{ mol H}}{1.008 \text{ g H}}\right) = 14.2 \text{ mol H}$

The empirical formula is CH_2. Its formula mass is 14 u.

$$\frac{84 \text{ u}}{14 \text{ u}} = 6$$

Hence, the molecular formula is C_6H_{12}.

75. In 100 g of compound there are

$(40.0 \text{ g C})\left(\dfrac{1 \text{ mol C}}{12.0 \text{ g C}}\right) = 3.33 \text{ mol C}$ $(6.7 \text{ g H})\left(\dfrac{1 \text{ mol H}}{1.0 \text{ g H}}\right) = 6.7 \text{ mol H}$

$(53.3 \text{ g O})\left(\dfrac{1 \text{ mol O}}{16.0 \text{ g O}}\right) = 3.33 \text{ mol O}$

to give a mole ratio

$$1 \text{ mol C} : 2 \text{ mol H} : 1 \text{ mol O}$$

The empirical formula is CH_2O. That formula has a formula mass of 30 u. There are thus

$$\frac{175 \text{ u}}{30 \text{ u}} = 6 \text{ formula units per molecule}$$

The molecular formula is therefore $C_6H_{12}O_6$.

Chapter 5: Background for the Structure of the Atom

76. $\dfrac{12.00 \text{ u}}{1 \text{ C atom}}\left(\dfrac{6.02 \times 10^{23} \text{ C atoms}}{1 \text{ mol C}}\right)\left(\dfrac{1 \text{ mol C}}{12.00 \text{ g}}\right) = \dfrac{6.02 \times 10^{23} \text{ u}}{1 \text{ g}}$

77. Atoms contain massive, positively charged centers (the nuclei).

78. 1.15×10^{-15}, the largest number that divides all the listed charges evenly. (The smallest of the charges listed is 2.30×10^{-15}, but this charge does not divide into all the others an integral number of times; hence 2.30×10^{-15} must represent the charge [in arbitrary units] of 2 electrons.)

79. Beta particles are much less massive, and therefore their charge-to-mass ratio is larger despite their lower charge.

80. **(A)** wave motion   **(B)** wave motion   **(C)** particles   **(D)** particles   **(E)** both   (In $E = h\nu$, E refers to the energy of each photon of light, while ν refers to the frequency of the waves of light.)

81. Violet, which has a shorter wavelength and thus a higher energy

$$E = h\nu = hc/\lambda$$

82. The light of 589 nm wavelength has precisely the energy absorbed by the sodium atom to promote its outermost electron from its ground state level to a higher level.

83. First line: $\dfrac{1}{\lambda} = 1.09678 \times 10^{7}\,\text{m}^{-1}\left(\dfrac{1}{1^{2}} - \dfrac{1}{2^{2}}\right) = 8.2259 \times 10^{6}\,\text{m}^{-1}$

$$\lambda = 1.2157 \times 10^{-7}\,\text{m}$$

Series limit: $\dfrac{1}{\lambda} = 1.09678 \times 10^{7}\,\text{m}^{-1}\left(\dfrac{1}{1^{2}} - \dfrac{1}{\infty}\right)$

$$= 1.09678 \times 10^{7}\,\text{m}^{-1}\left(\dfrac{1}{1^{2}} - 0\right) = 1.09678 \times 10^{7}\,\text{m}^{-1}$$

$$\lambda = 9.1176 \times 10^{-8}\,\text{m}$$

84. $\lambda = \dfrac{hc}{\Delta E} = \dfrac{(6.63 \times 10^{-34}\,\text{J}\cdot\text{s})(3.00 \times 10^{8}\,\text{m/s})}{4.4 \times 10^{-19}\,\text{J}} = 4.5 \times 10^{-7}\,\text{m}$

85. $5 \rightarrow 2$ (The lines of the Balmer series are first $3 \rightarrow 2$, second $4 \rightarrow 2$, third $5 \rightarrow 2$, fourth $6 \rightarrow 2$.)

86. $\dfrac{1}{\lambda} = R\left(\dfrac{1}{2^{2}} - \dfrac{1}{5^{2}}\right) = (1.09678 \times 10^{7}\,\text{m}^{-1})(0.2100) = 2.3032 \times 10^{6}\,\text{m}^{-1}$

$$\nu = \dfrac{c}{\lambda} = c\left(\dfrac{1}{\lambda}\right) = (3.00 \times 10^{8}\,\text{m/s})(2.3032 \times 10^{6}\,\text{m}^{-1}) = 6.91 \times 10^{14}\,\text{Hz}$$

The line is in the visible spectrum of hydrogen. It is a part of the Balmer series because the electron has had a transition to the second orbit from a higher orbit.

87. The emission of visible light implies some transition to the second orbit. After that, there *must* be a transition $2 \rightarrow 1$ for the atom to return to its ground state.

$$\dfrac{1}{\lambda} = 1.09678 \times 10^{7}\,\text{m}^{-1}\left(\dfrac{1}{1^{2}} - \dfrac{1}{2^{2}}\right) = 8.22585 \times 10^{6}\,\text{m}^{-1}$$

$$\lambda = 1.21568 \times 10^{-7}\,\text{m}$$

Chapter 6: Electronic Structure of the Atom

88. **(A)** l can be 0, 1, or 2.

(B) The value of m_l can vary from $-l$ to $+l$; in this case m_l can have values -2, -1, 0, $+1$, and $+2$.

(C) m_s must be $-\frac{1}{2}$ or $+\frac{1}{2}$. These are the only permitted values, regardless of the values of the other quantum numbers.

89. $n = 3$; $l = 1$; $m_l = -1$, 0, or 1; $m_s = +\frac{1}{2}$ or $-\frac{1}{2}$.

90. (A) $2n^2 = 2(2)^2 = 8$ **(B)** $2(3)^2 = 18$ **(C)** 8 (The $4s$ electrons precede the $3d$ electrons.)

91. $1s^2\, 2s^2\, 2p^6\, 3s^2\, 3p^3$

92. $1s^2\, 2s^2\, 2p^2$ group IVA

$1s^2\, 2s^2\, 2p^6\, 3s^2\, 3p^4$ group VIA

$1s^2\, 2s^2\, 2p^6\, 3s^2\, 3p^6\, 4s^2\, 3d^6$ group VIII

$1s^2\, 2s^2\, 2p^6\, 3s^2\, 3p^6\, 4s^2\, 3d^{10}\, 4p^6$ group 0

$1s^2\, 2s^2\, 2p^6\, 3s^2\, 3p^6\, 4s^2\, 3d^{10}\, 4p^6\, 5s^2\, 4d^{10}\, 5p^6\, 6s^2$ group IIA

93. Li $1s^2\, 2s^1$

    Na $1s^2\, 2s^2\, 2p^6\, 3s^1$

    K $1s^2\, 2s^2\, 2p^6\, 3s^2\, 3p^6\, 4s^1$

    Rb $1s^2\, 2s^2\, 2p^6\, 3s^2\, 3p^6\, 4s^2\, 3d^{10}\, 4p^6\, 5s^1$

    Fe $1s^2\, 2s^2\, 2p^6\, 3s^2\, 3p^6\, 4s^2\, 3d^6$

    F $1s^2\, 2s^2\, 2p^5$

94. The configurations of ions are derived from the configurations of their neutral atoms:

    S $1s^2\, 2s^2\, 2p^6\, 3s^3\, 3p^4$ S^{2-} $1s^2\, 2s^2\, 2p^6\, 3s^2\, 3p^6$

    Fe $1s^2\, 2s^2\, 2p^6\, 3s^2\, 3p^6\, 4s^2\, 3d^6$ Fe^{2+} $1s^2\, 2s^2\, 2p^6\, 3s^2\, 3p^6\, 4s^0\, 3d^6$

95. Br^- $1s^2\, 2s^2\, 2p^6\, 3s^2\, 3p^6\, 4s^2\, 3d^{10}\, 4p^6$

    Ca^{2+} $1s^2\, 2s^2\, 2p^6\, 3s^2\, 3p^6\, 4s^0$

    Fe^{2+} $1s^2\, 2s^2\, 2p^6\, 3s^2\, 3p^6\, 4s^0\, 3d^6$

    P^{3-} $1s^2\, 2s^2\, 2p^6\, 3s^2\, 3p^6$

96. Cl^- $1s^2\, 2s^2\, 2p^6\, 3s^2\, 3p^6$

    Ni^{2+} $1s^2\, 2s^2\, 2p^6\, 3s^2\, 3p^6\, 4s^0\, 3d^8$ $\dfrac{\text{↑↓ ↑↓ ↑↓ ↑ ↑}}{3d}$ so there are two unpaired electrons.

97. (A) Both are $1s^2\, 2s^2\, 2p^6\, 3s^2\, 3p^6$.

    (B) Fe: $1s^2\, 2s^2\, 2p^6\, 3s^2\, 3p^6\, 4s^2\, 3d^6$; Ni^{2+}: $1s^2\, 2s^2\, 2p^6\, 3s^2\, 3p^6\, 3d^8$

    (C) The first pair is isoelectronic. For the second, the last electrons added in Fe are $3d$, not the same as the $4s$ electrons removed on formation of Ni^{2+} ion.

    (D) Only the exceptional Pd is isoelectronic with Ag^+ and Cd^{2+}.

98. The Aufbau configuration for copper and the actual configuration for copper are as follows: $4s^2\, 3d^9$ and $4s^1\, 3d^{10}$. No matter which of these configurations is assumed to be the one on which the configuration of the ion is based, the removal of the $4s$ electron(s) before the $3d$ electrons results in the configuration for copper(II) ion: $4s^0$, $3d^9$.

99. **(A)** [Xe] $6s^2 5d^{10} 4f^{14}$ **(B)** [Xe] $6s^2 5d^{10} 4f^{14}$ **(C)** [Kr] $5s^2 4d^{10}$ **(D)** Each of these ions has retained the pair of electrons in its outermost s subshell—the inert pair. This pair of electrons is more difficult to remove than the p electrons, which have already been lost by these ions, but they are not totally unreactive. When lost, they form Pb^{4+}, Tl^{3+}, and Sn^{4+}, respectively.

100. Chemical and physical properties depend on electronic configuration; nuclear properties do not.

101. Their properties are similar because of similar outermost electronic configurations ($ns^2 np^5$). The differences stem from their differences in size and the magnitude of electronic shielding available for the outermost electrons.

102. Their outermost two shells of electrons are alike—$6s^2 5d^1$—or that configuration is very close to the ground state in stability. In any event, all the lanthanides act as if they had this configuration to a great extent.

103. **(A)** Ti and F⁻ **(B)** Na and Ne **(C)** Because of the lanthanide contraction (The increase in size in going from one period to the next is negated by the 14 extra decreases in size corresponding to the 14 "extra" lanthanide elements.)

104. Be^{2+}. Be is a second-period element and thus is smaller than higher period elements. The Be^{2+} ion is isoelectronic with Li^+ but has a greater nuclear charge.

105. Na. The size decreases as one goes to the right in the periodic table, and also decreases as one removes electrons.

106. P and Cl are members of the same period. Cl should have a smaller radius in keeping with the usual trend across a period. The experimental value is 100 pm.

107. The IE of Na should be close to the arithmetic average of the two, or 4.9 eV. (The observed value is 5.1 eV.)

108. Sr. It is a metal.

109. The large differences in successive ionization energies occur at the points when electrons are first removed from completed s plus p subshells (the "octet").

110. Copper has 10 more protons and 10 more electrons than does K, but the electrons do not screen the nucleus perfectly. Hence, the first ionization energy of Cu is higher. The second ionization energy for K involves the loss of an electron from an octet—an inner, complete s plus p subshell arrangement, whereas that for Cu involves the more easily ionized d^{10} configuration.

111. Ce^{4+} has the stable electronic configuration of the rare gas Xe. Eu^{2+}, with 61 electrons, has the configuration [Xe] $4f^7$, with the added stability of a half-filled $4f$ subshell.

Chapter 7: Bonding

112. Covalent-bond radii from Table 7.1 are merely added. Be sure that data for the correct bond orders are used.

(A) 70 pm + 28 pm = 98 pm   **(B)** 104 pm + 99 pm = 203 pm

(C) C—H   77 pm + 28 pm = 105 pm   C—Cl   77 pm + 99 pm = 176 pm

113. **(A)** C — C < C $=$ C < C $\equiv$ C   **(B)** C $\equiv$ C < C $=$ C < C — C

114. Internuclear distance – radius of chlorine atom = radius of As atom:

$$220 \text{ pm} - 99 \text{ pm} = 121 \text{ pm}$$

115. In compound A, the carbon–oxygen distance is the sum of the single-bond covalent radii of carbon and oxygen:

$$77 \text{ pm} + 66 \text{ pm} = 143 \text{ pm}$$

The oxygen must therefore be bonded to two carbon atoms. One such structure, the actual one selected for this experiment, is the heterocyclic compound tetrahydrofuran.

In compound B, the carbon–oxygen distance is close to that predicted for a C$=$O double bond, 122 pm, such as in butanone.

116. H—C$\equiv$C—H   The length of the molecule is the sum of two hydrogen atom radii, two carbon single-bond radii, and two carbon triple-bond radii:

$$2(28 \text{ pm}) + 2(77 \text{ pm}) + 2(61 \text{ pm}) = 332 \text{ pm}$$

117. Covalent bonds between third-period nonmetals are generally weaker than bonds between atoms of their second-period neighbors. Si—Si bonds are expected to be weaker than C—C bonds; the bond strength should be less than 300 kJ/mol.

118. Electronegativity is only semiquantitative, and therefore the variations from element to element are less detailed. Moreover, electronegativity for some noble gas elements is undefined.

119. Note that single-bond energies are used.

$$\Delta = D(N - F) - \sqrt{D(N - N)D(F - F)}$$
$$= 225 \text{ kJ} - \sqrt{(128 \text{ kJ})(151 \text{ kJ})} = 86 \text{ kJ/mol}$$
$$EN(N) = EN(F) - 0.102\sqrt{\Delta} = 3.05$$

The tabulated value of 3.0 for nitrogen is an average value from many compounds.

120. A polar bond is formed between two atoms of different electronegativity. A polar molecule results when there is only one polar bond or when the polar bonds within it are not symmetrically oriented to cancel the effects of the others. The polar molecule is often referred to as having a dipole moment.

121. Both of these molecules have the same number of electrons and almost the same molecular masses, but ICl has a dipole moment. Therefore, ICl is expected to have the higher boiling point. At 1 atm pressure, the measured boiling points are ICl, 97.4°C, and Br_2, 58.78°C.

122. $BF_3 < H_2S < H_2O$ BF_3 has a zero dipole moment because of its symmetry. H_2S has a lower dipole moment than H_2O because of the much lower bond polarity of H—S compared with H—O.

123. (A) No.   **(B)** If the various polar bonds are arranged symmetrically, as in CCl_4 and CO_2, the effects of one bond are canceled by the effects of the other(s).

124. van der Waals < dipole < hydrogen bonding < covalent bond

125. High melting point, high boiling point, high heat of fusion, high heat of vaporization, low density of ice compared with water, high specific heat, high ionic conductance of hydronium and hydroxide ions, and many others.

126. The greater the atomic number, the greater the number of electrons in each atom, and the greater is the van der Waals force. The greater the forces between the molecules, the higher should be the boiling point. The actual boiling points of the noble gases increase with increasing atomic number, as expected: He, 4 K; Ne, 27 K; Ar, 87 K; Kr, 120 K; Xe, 166 K; Rn, 211 K.

127. (A) CH_3CH_2OH (H is connected to O)   **(B)** CH_3NH_2 (S does not form hydrogen bonds)

128. NH_3 has the strongest intermolecular forces; thus, it is expected to have the highest melting point. (Actual melting points are NH_3, −77.7°C, PH_3, −133°C; $(CH_3)_3N$, −117°C.)

129.

130. (A)

(B) This is an actual chemical reaction (an equilibrium). Resonance forms are only multiple representations of the same structure.

131. The forms shown as **(D)** and **(E)** cannot be used. In **(D)** the atoms are arranged with the oxygen atom between the nitrogen atoms. This arrangement is not the same as in the other forms or as the position found experimentally. The form shown as **(E)** has four unpaired electrons, and this is not possible, since the molecule is diamagnetic.

132. (A) trigonal bipyramidal, no **(B)** angular, yes **(C)** linear, no **(D)** trigonal planar, no **(E)** angular, yes

133. Iodine has an unshared electron pair. This unshared pair must be assigned a region of space sufficiently removed from the I—F bonds, to minimize electron repulsion. The compact structure of the trigonal bipyramid does not allow room for the unshared pair. The square pyramid structure of IF_5 may be thought of as an octahedron with the unshared pair pointing to one of the corners.

134. (A) $\left[\begin{array}{c} \text{H} \\ \text{H}:\ddot{\text{O}}:\text{H} \end{array} \right]^{+}$ trigonal pyramidal

(B) $\begin{array}{c} \text{H.} \qquad \text{.H} \\ :\text{C}::\text{N}: \\ \text{H} \end{array}$ planar [The carbon atom is trigonal (planar); the nitrogen atom is angular. The whole molecule lies in one plane.]

(C) $:\ddot{\text{O}}:\ddot{\text{Cl}}:\ddot{\text{O}}:^{-}$ angular

(D) $\left[\begin{array}{c} \text{H} \\ \text{H}:\ddot{\text{N}}:\text{H} \\ \text{H} \end{array} \right]^{+}$ tetrahedral

(E) $\begin{array}{c} \text{H}:\ddot{\text{N}}:\ddot{\text{N}}:\text{H} \\ \text{H} \quad \text{H} \end{array}$ Both nitrogen atoms are pyramidal; the molecule is nonplanar.

135. (A) trigonal planar **(B)** trigonal pyramidal **(C)** trigonal planar **(D)** tetrahedral **(E)** trigonal pyramidal

136. (A) NH_3 (Lone pairs expand more, crowding bonding pairs together. NH_3 has only one lone pair; H_2O has two.) **(B)** BeF_2 (*sp* hybrids are linear.) **(C)** BF_3 **(D)** NH_3 **(E)** NH_3

137. Trigonal bipyramidal structures have inherently two different types of bonding—axial, 90° away from a plane, and equatorial, in the plane. The equatorial bonds are 120° away from each other. Since the bonding is different, the bond lengths are not expected to be exactly the same. SF_6 is octahedral; all 6 bonds are exactly the same.

Chapter 8: Bonding Theory

138. The *p* orbital has equal-sized lobes; most of the *sp* probability density is on one side, making the latter more directional in character.

139. (A) square planar and trigonal planar, respectively. **(B)** *sp*, sp^3d^2 or d^2sp^3, and dsp^3, respectively.

140. (A) The ground-state carbon atom configuration has two electrons in each of the 2s and 2p subshells. One electron is promoted to increase the number of unpaired electrons in the s and p subshells to the maximum, in this case 4. Since two atoms are bonded to each carbon atom, which has no lone pairs, 2 hybrid orbitals are required. The hybrid orbitals formed are *sp*, which are used to form σ bonds to the bonded atoms. The other two unpaired electrons on each atom pair up in π bonds with the similar electrons on the other carbon atom. A linear molecule results. **(B)** The sulfur atom has a ground state with two electrons in its $3s$ subshell and four in its $3p$ subshell. Since 6 uncharged fluorine atoms are to be added, 6 unpaired electrons are required in 6 hybrid orbitals. The sp^3d^2 hybrids that result cause formation of an octahedral molecule.

141. **(A)** sp, linear   **(B)** sp^3, tetrahedral   **(C)** sp^2, planar   **(D)** sp^3d^2, octahedral
(E) sp^2, trigonal planar   **(F)** sp, linear

142. **(A)** sp^3, pyramidal   **(B)** sp^2, planar (trigonal about each C)

(C) sp^3, pyramidal: O↿⇂ O↿⇂ O↿⇂ ↿⇂

143. A σ orbital has greater electron overlap between the atoms because its component p atomic orbitals are directed toward each other, whereas the component p orbitals making up the π orbital are directed perpendicular to the internuclear axis.

144. C_2H_4 has a π bond. The overlap of the atomic p orbitals making up this bond would be destroyed if the —CH$_2$ groups did rotate.

145. **(A)** AX$_5$ is the only molecule with 5 atoms bonded to a central atom—it is the only possible molecule listed that could be trigonal bipyramidal. **(B)** In the absence of lone pairs, it must be dsp^3 or sp^3d. (An inner or a valence d orbital may be involved.)

146. **(A)** It has four lobes positioned between the y and x axes. **(B)** It has one large cylindrically symmetric lobe along the x axis between the two nuclei. **(C)** Four lobes are situated above and below the x axis on the sides away from the farther nucleus. (At those positions, electron density will not tend to pull the two atoms together.) **(D)** It has a large lobe along the x axis pointed toward the atom that it bonds.

147. **(A)**   — σ^*_{2p}

   —   — π^*_{2p}

   — σ_{2p}

   —   — π_{2p}

(B) Eight electrons fill the σ_{1s}, σ^*_{1s}, σ_{2s}, σ^*_{2s} orbitals. The other six $2p$ electrons of the two atoms fill the lowest 3 orbitals shown, creating a triply bonded species: $\pi^4_{2p}\,\sigma^2_{2p}$

(C) The CN$^-$ ion is isoelectronic with CO.

148. The electronic configuration of NO can be drawn by adding the three $2p$ electrons from N and the four from O to the diagram in Question 147. The 6 bonding and 1 anti-bonding electrons yield a net of 5 bonding electrons, for a bond order of $2\frac{1}{2}$.

149. NO$^+$ has lost an antibonding electron; CO$^+$ has lost a bonding electron.

150. O_2^{2-} has a bond order of 1; O_2 has a bond order of 2. Hence, O_2^{2-} will have a longer (and weaker) bond.

151. Nonbonding orbitals have the same energy as the atomic orbitals from which they are formed; antibonding orbitals have higher energies than the highest energy atomic orbital from which they are formed.

152. Neither   (Use the energy level diagram of Question 147.)

153. $(p_x, p_x), (p_x, d_{xz}), (d_{xz}, d_{xz})$

154. The σ_{2s} orbital originates from atomic orbitals of higher energy than itself, and so is bonding. The σ_{1s}^* originates from atomic orbitals of lower energy than itself, and so is antibonding. The relative energies of the two σ orbitals is immaterial.

Chapter 9: Organic Molecules

155. (A) 2,3-dimethylbutane   **(B)** and **(C)** 2,2-dimethylbutane

156. (A) $CH_3 \!-\! CH \!-\! CH \!-\! CH_2 \!-\! CH_2 \!-\! CH_3$
with CH_3 and $CH_2 \!-\! CH_3$ substituents

(B) $CH_3 \!-\! \underset{\underset{CH_3}{|}}{\overset{\overset{CH_3}{|}}{C}} \!-\! CH_2 \!-\! CH_2 \!-\! CH_2 \!-\! CH_3$

(C) $CH_3 \!-\! CH \!-\! CH \!-\! CH_2 \!-\! CH_2 \!-\! CH \!-\! CH_3$
with CH_3, CH_3 and CH_3 substituents

157. Both **(A)** and **(D)** have 5-carbon continuous chains and should be called pentane. **(B)** and **(C)** are both the same, and no number is needed. It is methylbutane.

158. (A) 1   **(B)** 4   **(C)** 1   **(D)** 2   **(E)** 3   (except in ionic compounds)

159. 2-methyl-3-ethylheptane   (starting at either CH_3 group on the bottom row)

160. Ethers. CH_3OCH_3, $CH_3OC_6H_5$, $C_6H_5OC_6H_5$. (The fact that the radicals are denoted R and R′ means that they *may* be different, not that they *must* be different.)

161. (A) 2-methyl-1-butanol   **(B)** 3-methyl-1-butanol   **(C)** 3-hexanol

(D) 4-methyl-1-pentene   **(E)** 2-pentene

162. (A) 4-chloro-2-methyl-1-butene   **(B)** octanoic acid   **(C)** 2-methylpentanal

163. (A) Butanol may have the OH group on the first or second carbon atom. Butanone cannot have the carbonyl group on the end (where it would be an aldehyde functional group).   **(B)** Pentanone could refer to 2-pentanone or to 3-pentanone, so pentanone is not specific.

164. (A) $CH_3CH_2NH_2$ (aminoethane)   **(B)** CH_3CH_2CHO (propanal)

(C) $CH_3COCH_2CH_3$ (butanone)   **(D)** $CH_3CH_2CO_2CH_2CH_3$ (ethyl propanoate)

(E) $HCO_2CH_2CH_2CH_2CH_3$ (butyl methanoate)   **(F)** C_6H_5Br (bromobenzene)

(G) $HC \equiv CH$ (ethyne)   **(H)** $C_6H_5C \equiv CH$ (phenylethyne)

165. (A) $CH_3CH = CHCH_3$ **(B)** $CH_3OCH_2CH_3$ **(C)** CH_3CH_2CHO

(D) $CH_3CHOHCH_3$

166. (A) $CH_3CH_2CH_2CHCH_2CH_2CH_3$ **(B)** $CH_3CH_2CH_2CHCH_2CH_2CH_3$

$\qquad\qquad\quad CH_2 - CH_3 \qquad\qquad\qquad\qquad\quad CH_2 - CH_2 - CH_3$

167. (A) amines   **(B)** carboxylic acids

168.

Halide	CH_3F	fluoromethane (methyl fluoride)
Alcohol	CH_3OH	methanol (methyl alcohol)
Ether	$(CH_3)_2O$	dimethyl ether
Aldehyde	CH_2O	methanal (formaldehyde)
Ketone	CH_3COCH_3	propanone (acetone)
Acid	HCO_2H	methanoic acid (formic acid)
Ester	HCO_2CH_3	methyl methanoate (methyl formate)
Amine	CH_3NH_2	aminomethane (methyl amine)
Amide	$HCONH_2$	methanamide (formamide)
Alkene	$H_2C = CH_2$	ethene (ethylene)
Alkyne	$HC \equiv CH$	ethyne (acetylene)
Aromatic	C_6H_6	benzene

169. (A) ether, methyl ethyl ether   **(B)** acid, propanoic acid   **(C)** amide, ethanamide
(D) ester, methyl ethanoate

170. Nonane

171. $CH_2 = CHCH_2CH_3$ $CH_3CH = CHCH_3$ $H_2C = CCH_3$ $H_2C - CHCH_3$
$$\qquad\qquad\qquad\qquad\qquad\qquad\qquad\qquad\qquad\quad |\qquad\quad\diagdown\diagup$$
$$\qquad\qquad\qquad\qquad\qquad\qquad\qquad\qquad\quad CH_3\qquad CH_2$$

172. There are 7: 1-butanol, 2-butanol, 2-methyl-1-propanol, 2-methyl-2-propanol, diethyl ether, methyl 1-propyl ether, and methyl 2-propyl ether.

173. $CH_3CH_2CH_2CH_2CH_3$ $CH_3CH_2CHCH_3$ CH_3CCH_3
$$\qquad\qquad\qquad\qquad\qquad\qquad\qquad\qquad\qquad\quad |\qquad\qquad\qquad\overset{\displaystyle CH_3}{\underset{\displaystyle CH_3}{|}}$$
$$\qquad\qquad\qquad\qquad\qquad\qquad\qquad\qquad\quad CH_3$$

174. 1,1-diiodobutane, 1,2-diiodobutane, 1,3-diiodobutane, 1,4-diiodobutane, 2,2-diiodobutane, 2,3-diiodobutane, 1,1-diiodo-2-methyl-propane, 1,2-diiodo-2-methyl-propane, 1,3-diiodo-2-methyl-propane

175. $CH_2 = CHCH_2CH_2CH_3$ $CH_3CH = CHCH_2CH_3$ $CH_2 = CHCH(CH_3)_2$

$CH_3CH = C(CH_3)_2$ $CH_3CH_2(CH_3)C = CH_2$

$$H_2C - CH_2 \qquad H_2C - CH - CH_3 \qquad H_2C - CH - CH_2CH_3$$
$$\diagup \qquad\quad \diagdown \qquad\quad |\qquad | \qquad\qquad \diagdown\diagup$$
$$H_2C \qquad\quad CH_2 \qquad H_2C - CH_2 \qquad\qquad CH_2$$
$$\diagdown \quad\diagup$$
$$CH_2$$

$$H_3C - CH - CH - CH_3$$
$$\diagdown\diagup$$
$$CH_2$$

176. If the four carbons are not in a ring, there must be one double bond to satisfy the tetracovalence of carbon. The double bond occurs either in the center of the molecule or toward an end. In the former case, two geometric isomers occur with different positions of the terminal carbons relative to the double bond; in the latter case, two structural isomers occur, differing in the extent of branching within the skeleton. Additional possibilities are ring structures (without double bonds).

$$H_3C\diagdown\qquad\diagup CH_3 \qquad H_3C\diagdown\qquad\diagup H \qquad H_3C\diagdown\qquad\diagup H$$
$$\qquad C = C \qquad\qquad\qquad C = C \qquad\qquad\qquad C = C$$
$$H\diagup\qquad\diagdown H \qquad\quad H\diagup\qquad\diagdown CH_3 \qquad H_3C\diagup\qquad\diagdown H$$

$$CH_3CH_2\diagdown\qquad\diagup H \qquad\quad H_2C - CH_2 \qquad\qquad CH_2$$
$$\qquad C = C \qquad\qquad\qquad |\qquad | \qquad\qquad\quad \diagup\quad\diagdown$$
$$H\diagup\qquad\diagdown H \qquad\quad H_2C - CH_2 \qquad H_2C - CH - CH_3$$

177. Only CHBr=CHCl can exist as geometric isomers:

$$\underset{H}{\overset{Br}{>}}C=C\underset{H}{\overset{Cl}{<}} \quad \text{and} \quad \underset{H}{\overset{Br}{>}}C=C\underset{Cl}{\overset{H}{<}}$$

In CH_2Cl-CH_2Cl and CH_2Cl-CH_2Br, the carbon atoms are connected by a single bond about which the groups can rotate relatively freely. Thus, any conformation of the halogen atoms may be converted into any other simply by rotation about the single bond. In CH_2Cl_2, the configuration of the molecule is tetrahedral, and all interchanges of atoms yield exactly equivalent configurations.

178. The first and third compound. (The second Compound has both halogen atoms on the same carbon atom.)

Chapter 10: Chemical Equations

179. The absence implies a coefficient of 1. Before the equation is completely balanced, however, the lack of a coefficient might merely signify that the species has not yet been done. Take care to distinguish between these meanings.

180. There are no fixed rules for balancing simple equations. Often a trial-and-error procedure is used. It is commonly helpful to start with the most complex formula. In this case, Fe_2O_3 has two different elements and a greater total number of atoms than any of the other substances, so we might start with it.

$$FeS_2 + O_2 \rightarrow 1\,Fe_2O_3 + SO_2$$

Balance the iron atoms: $\quad 2\,FeS_2 + O_2 \rightarrow 1\,Fe_2O_3 + SO_2$

Balance sulfur: $\quad 2\,FeS_2 + O_2 \rightarrow 1\,Fe_2O_3 + 4\,SO_2$

Balance oxygen: $\quad 2\,FeS_2 + \frac{11}{2}O_2 \rightarrow 1\,Fe_2O_3 + 4\,SO_2$

Clear the fraction by multiplying *every* coefficient by 2:

$$4\,FeS_2 + 11\,O_2 \rightarrow 2\,Fe_2O_3 + 8\,SO_2$$

181. First write the correct formulas, and only then balance the equation!

$$2\,ZnS + 3\,O_2 \rightarrow 2\,ZnO + 2\,SO_2$$
$$2\,HNO_3 + CuCO_3 \rightarrow H_2O + CO_2 + Cu(NO_3)_2$$

182. **(A)** $4\,BCl_3 + P_4 + 6\,H_2 \rightarrow 4\,BP + 12\,HCl$

(B) $2\,C_2H_2Cl_4 + Ca(OH)_2 \rightarrow 2\,C_2HCl_3 + CaCl_2 + 2\,H_2O$

(C) $(NH_4)_2Cr_2O_7 \rightarrow N_2 + Cr_2O_3 + 4\,H_2O$

(D) $2\,C_8H_{18} + 17\,O_2 \rightarrow 16\,CO + 18\,H_2O$

183. (A) $NCl_3 + 3H_2O \rightarrow NH_3 + 3HOCl$

(B) $PCl_3 + 3H_2O \rightarrow H_3PO_3 + 3HCl$

(C) $SbCl_3 + H_2O \rightarrow Sb(O)Cl + 2HCl$

184. 1. Combination reactions, e.g., $2Na + Cl_2 \rightarrow 2NaCl$
2. Decomposition reactions, e.g., $2HgO \rightarrow 2Hg + O_2$
3. Replacement reactions, e.g., $2NaI + Cl_2 \rightarrow 2NaCl + I_2$
4. Double replacement reactions, e.g., $AgNO_3 + NaCl \rightarrow AgCl + NaNO_3$
5. Combustion reactions, e.g., $CH_4 + 2O_2 \rightarrow CO_2 + 2H_2O$

185. (A) combination   **(B)** double replacement   **(C)** decomposition   **(D)** replacement
(E) double replacement

186. (A) acid   **(B)** basic anhydride   **(C)** base   **(D)** basic anhydride   **(E)** acid anhydride
(F) basic anhydride   **(G)** acid anhydride   **(H)** basic anhydride

187. (A) replacement reactions   **(B)** double replacement reactions

188. (A) acetates, nitrates, chlorates, alkali metal ions, and ammonium ion
(B) $AgCl$, $PbCl_2$, $CuCl$, and Hg_2Cl_2

189. (A) $2H_2O_2 \xrightarrow{\text{heat}} 2H_2O + O_2$ decomposition

(B) $2H_2 + O_2 \xrightarrow{\text{spark}} 2H_2O$ combination

(C) $2Na + Cl_2 \rightarrow 2NaCl$ combination

(D) $FeCl_2 + 2AgNO_3 \rightarrow Fe(NO_3)_2 + 2AgCl$ double replacement

(E) $CaO + CO_2 \rightarrow CaCO_3$ combination

190. (A) $Zn + FeCl_2 \rightarrow ZnCl_2 + Fe$

(B) $F_2 + 2NaCl \rightarrow 2NaF + Cl_2$

(C) $NaCl + AgNO_3 \rightarrow AgCl + NaNO_3$

(D) $AgCl + NaNO_3 \rightarrow$ no reaction   [Compare part **(C)**.]

(E) $H_3PO_4 + NaOH \text{ (limited)} \rightarrow NaH_2PO_4 + H_2O$

(F) $KHSO_4 + KOH \rightarrow K_2SO_4 + H_2O$

(G) $SO_2 + H_2O \rightarrow H_2SO_3$

(H) $P_2O_3 + 3H_2O \rightarrow 2H_3PO_3$

191. (A) $2\,AgNO_3 + Cu \rightarrow Cu(NO_3)_2 + 2\,Ag$ **(E)** $2\,HCl + Ba(OH)_2 \rightarrow BaCl_2 + 2\,H_2O$

(B) $3\,Cl_2 + 2\,Al \rightarrow 2\,AlCl_3$ **(F)** $BaCl_2 + K_2SO_4 \rightarrow BaSO_4 + 2\,KCl$

(C) $2\,NaI + Cl_2 \rightarrow 2\,NaCl + I_2$ **(G)** $2\,KClO_3 \xrightarrow{\text{heat}} 2\,KCl + 3\,O_2$

(D) $2\,CO + O_2 \rightarrow 2\,CO_2$

192. (A) $NH_4Cl + NaOH \rightarrow NaCl + NH_3 + H_2O$

(Un-ionized products are formed from ionic reactants. NH_4OH is unstable and decomposes to NH_3 and H_2O.)

(B) $NaC_2H_3O_2 + HCl \rightarrow NaCl + HC_2H_3O_2$

(C) $Ca + 2\,H_2O \rightarrow Ca(OH)_2 + H_2$

(D) $4\,Li + O_2 \rightarrow 2\,Li_2O$

(E) $Mg + 2\,HCl \rightarrow MgCl_2 + H_2$

(F) $2\,C_6H_6 + 15\,O_2 \rightarrow 12\,CO_2 + 6\,H_2O$ or $C_6H_6 + 7\tfrac{1}{2}O_2 \rightarrow 6\,CO_2 + 3\,H_2O$

(G) no reaction

193. Ions are free to move relatively free from the influence of other ions only in solution. There are few other solvents that dissolve ions well.

194. Sodium forms very few compounds that are insoluble in water, and it forms no gaseous compounds at ordinary temperatures. (Sodium forms only ionic compounds.)

195. Only ions in solution that appear unchanged on both sides of the equation are omitted in the net ionic equation. They are called spectator ions. If an ion changes in any way— for example, changes into part of a solid or part of a covalent compound or changes its charge—it appears as the ion on at least one side.

196. (A) Ag_2S is insoluble, while $NaClO_3$ is soluble.

$$2\,Ag^+ + 2\,\cancel{ClO_3^-} + 2\,\cancel{Na^+} + S^{2-} \rightarrow Ag_2S(s) + 2\,\cancel{Na^+} + 2\,\cancel{ClO_3^-}$$

Net ionic equation: $2\,Ag^+ + S^{2-} \rightarrow Ag_2S(s)$

(B) $Hg_3(PO_4)_2$ is insoluble, while $(NH_4)_2SO_4$ is soluble.

Net ionic equation:

$$6\,\cancel{NH_4^+} + 2\,PO_4^{3-} + 3\,Hg^{2+} + 3\,\cancel{SO_4^{2-}} \rightarrow 6\,\cancel{NH_4^+} + 3\,\cancel{SO_4^{2-}} + Hg_3(PO_4)_2(s)$$

$$3\,Hg^{2+} + 2\,PO_4^{3-} \rightarrow Hg_3(PO_4)_2(s)$$

197. (A) $SO_4^{2-} + Ba^{2+} \rightarrow BaSO_4$

(B) $H^+ + HCO_3^- \rightarrow CO_2 + H_2O$

(C) $2H^+ + Fe(OH)_2(s) \rightarrow Fe^{2+} + 2H_2O$

(D) $H^+ + HCO_3^- \rightarrow CO_2 + H_2O$

(E) $2Fe^{3+} + Fe \rightarrow 3Fe^{2+}$

(F) $OH^- + NH_4^+ \rightarrow NH_3 + H_2O$

198. (A) $HCO_3^- + OH^- \rightarrow CO_3^{2-} + H_2O$ **(D)** $H^+ + OH^- \rightarrow H_2O$

(B) $Zn + Hg_2^{2+} \rightarrow Zn^{2+} + 2Hg$ **(E)** $Zn + 2H^+ \rightarrow Zn^{2+} + H_2$

(C) $2Ag^+ + H_2S \rightarrow Ag_2S + 2H^+$ **(F)** $Cu^{2+} + H_2S \rightarrow CuS + 2H^+$

Chapter 11: Stoichiometry

199. $(2.50 \text{ mol } CO_2)\left(\dfrac{1 \text{ mol } Ca(HCO_3)_2}{2 \text{ mol } CO_2}\right) = 1.25 \text{ mol } Ca(HCO_3)_2$

200. $Ba(OH)_2 + CO_2 \rightarrow BaCO_3 + H_2O$

$(0.205 \text{ mol } Ba(OH)_2)\left(\dfrac{1 \text{ mol } BaCO_3}{\text{mol } Ba(OH)_2}\right)\left(\dfrac{197.4 \text{ g } BaCO_3}{\text{mol } BaCO_3}\right) = 40.5 \text{ g}$

201. (A) $1.00 \text{ g sucrose}\left(\dfrac{1 \text{ mol sucrose}}{342 \text{ g sucrose}}\right)\left(\dfrac{12 \text{ mol } CO_2}{1 \text{ mol sucrose}}\right) = 0.0351 \text{ mol } CO_2$

$(0.0351 \text{ mol } CO_2)\left(\dfrac{44.0 \text{ g } CO_2}{\text{mol } CO_2}\right) = 1.54 \text{ g } CO_2$

(B) According to the balanced chemical equation, for every mole of CO_2 produced 1 mol of O_2 is needed, and in this case 0.0351 mol of O_2 will be used up.

202. $2KClO_3 \rightarrow 2KCl + 3O_2$

(A) $(1.23 \text{ g } O_2)\left(\dfrac{1 \text{ mol } O_2}{32.0 \text{ g } O_2}\right)\left(\dfrac{2 \text{ mol } KClO_3}{3 \text{ mol } O_2}\right)\left(\dfrac{122.6 \text{ g } KClO_3}{\text{mol } KClO_3}\right) = 3.14 \text{ g } KClO_3$

(B) $(3.14 \text{ g } KClO_3) - (1.23 \text{ g } O_2) = 1.91 \text{ g } KCl$

203. $(1000 \text{ g } I_2)\left(\dfrac{1 \text{ mol } I_2}{253.8 \text{ g } I_2}\right) = 3.94 \text{ mol } I_2$

$(3.94 \text{ mol } I_2)\left(\dfrac{2 \text{ mol } NaIO_3}{\text{mol } I_2}\right)\left(\dfrac{197.9 \text{ g } NaIO_3}{\text{mol } NaIO_3}\right) = 1.56 \times 10^3 \text{ g} = 1.56 \text{ kg } NaIO_3$

$(3.94 \text{ mol } I_2)\left(\dfrac{5 \text{ mol } NaHSO_3}{\text{mol } I_2}\right)\left(\dfrac{104 \text{ g } NaHSO_3}{\text{mol } NaHSO_3}\right) = 2.05 \times 10^3 \text{ g} = 2.05 \text{ kg } NaHSO_3$

204. (A) $Cu_2S + O_2 \rightarrow 2\,Cu + SO_2$

(B) $(501 \text{ g } Cu_2S)\left(\dfrac{1 \text{ mol } Cu_2S}{159 \text{ g } Cu_2S}\right)\left(\dfrac{2 \text{ mol } Cu}{\text{mol } Cu_2S}\right)\left(\dfrac{63.5 \text{ g } Cu}{\text{mol } Cu}\right) = 400 \text{ g } Cu$

205. $Na_2SO_3 + 2\,HCl \rightarrow SO_2 + 2\,NaCl + H_2O$

$(105 \text{ g } Na_2SO_3)\left(\dfrac{1 \text{ mol } Na_2SO_3}{126 \text{ g } Na_2SO_3}\right)\left(\dfrac{1 \text{ mol } SO_2}{1 \text{ mol } Na_2SO_3}\right)\left(\dfrac{64.1 \text{ g } SO_2}{\text{mol } SO_2}\right) = 53.4 \text{ g } SO_2$

206. $CaCO_3 + 2\,HCl \rightarrow CaCl_2 + CO_2 + H_2O$

$0.880 \text{ g } CO_2\left(\dfrac{1 \text{ mol } CO_2}{44.0 \text{ g } CO_2}\right)\left(\dfrac{1 \text{ mol } CaCO_3}{1 \text{ mol } CO_2}\right) = 0.0200 \text{ mol } CaCO_3$

$(0.0200 \text{ mol } CaCO_3)\left(\dfrac{100 \text{ g } CaCO_3}{\text{mol } CaCO_3}\right) = 2.00 \text{ g } CaCO_3$

$\% \text{ } CaCO_3 \text{ in mixture} = \dfrac{2.00 \text{ g } CaCO_3}{4.00 \text{ g mixture}} \times 100\% = 50.0\%$

207. The masses of C and H in the compound are calculated from the masses of the products:

$(23.1 \text{ mg } CO_2)\left(\dfrac{1 \text{ mmol } CO_2}{44.0 \text{ mg } CO_2}\right)\left(\dfrac{1 \text{ mmol } C}{\text{mmol } CO_2}\right) = 0.525 \text{ mmol } C$

$(0.525 \text{ mmol } C)\left(\dfrac{12.0 \text{ mg } C}{\text{mmol } C}\right) = 6.30 \text{ mg } C$

$(4.72 \text{ mg } H_2O)\left(\dfrac{1 \text{ mmol } H_2O}{18.0 \text{ mg } H_2O}\right)\left(\dfrac{2 \text{ mmol } H}{\text{mmol } H_2O}\right) = 0.524 \text{ mmol } H$

$(0.524 \text{ mmol } H)\left(\dfrac{1.008 \text{ mg } H}{\text{mmol } H}\right) = 0.528 \text{ mg } H$

The mass of oxygen in the unknown compound is determined by difference:

$(10.20 \text{ mg}) - (6.30 \text{ mg}) - (0.528 \text{ mg}) = 3.37 \text{ mg } O$

$(3.37 \text{ mg } O)\left(\dfrac{1 \text{ mmol } O}{16.0 \text{ mg } O}\right) = 0.211 \text{ mmol } O$

The mole ratio is (0.525 mmol C):(0.524 mmol H):(0.211 mmol O) or (5 mol C):(5 mol H):(2 mol O). The empirical formula is $C_5H_5O_2$.

208. Using the balanced chemical equation, determine the ratio of moles of the two reactants present and that of moles required by the reaction:

Present

$$\frac{1.50 \text{ mol A}}{0.50 \text{ mol B}} = 3.0$$

Required

$$\frac{3 \text{ mol A}}{2 \text{ mol B}} = 1.5$$

Since there is more A present per mol of B than is required, B is limiting and will be totally used up in the reaction.

All quantities are in moles:

	3 A	+	2 B	→	C	+	3 D
Present initially:	1.50		0.50		0		0
Change due to reaction:	− 0.75		− 0.50		+ 0.25		+ 0.75
Present finally:	0.75		0.00		0.25		0.75

Note that the quantities in the row marked "Change due to reaction" are in the same ratio as the balanced chemical equation, but in the other rows they are not.

209. We know that O_2 is limiting; if it were not, at least some CO_2 would be produced (or the question is faulty).

	2 C	+	O_2	→	2 CO
Initial:	0.179 mol		0.0810 mol		0.000 mol
Change:	− 0.162 mol		− 0.0810 mol		+ 0.162 mol
Final:	0.017 mol		0.0000 mol		0.162 mol

210. Change the given mass to moles:

$$17.0 \text{ g NaHCO}_3 \left(\frac{1 \text{ mol NaHCO}_3}{84.0 \text{ g NaHCO}_3} \right) = 0.202 \text{ mol NaHCO}_3 \text{ present}$$

	2 NaHCO$_3$	+	H$_2$SO$_4$	→	Na$_2$SO$_4$	+	2 CO$_2$	+	2 H$_2$O
Initial:	0.202		0.400		0.000				
Change:	− 0.202		− 0.101		+ 0.101				
Final:	0.000		0.299		0.101				

Change the number of moles of Na$_2$SO$_4$ to grams:

$$0.101 \text{ mol Na}_2\text{SO}_4 \left(\frac{142 \text{ g Na}_2\text{SO}_4}{1 \text{ mol Na}_2\text{SO}_4} \right) = 14.3 \text{ g Na}_2\text{SO}_4$$

211. $PCl_5 + 4H_2O \rightarrow H_3PO_4 + 5HCl$

From these quantities, we can guess that H_2O is in excess. (If we guess incorrectly, a negative quantity of a reactant will upper in the "Final" row, and we can go back and redo the calculation. This procedure usually saves a step in the solution.)

	PCl_5	$+$	$4H_2O$	$\rightarrow$	H_3PO_4	$+$	$5H_2O$
Initial:	0.250		2.00		0.00		0.00
Change:	-0.250		-1.00		$+0.250$		$+1.25$
Final:	0.000		1.00		0.250		1.25

Note that the ratio of the magnitudes of PCl_5 to H_2O to H_3PO_4 to HCl in the "Change" row is $1 : 4 : 1 : 5$, just as in the balanced chemical equation.

212. Seeing that $AgNO_3$ is limiting, we get:

	$BaCl_2(aq)$	$+$	$2\,AgNO_3(aq)$	$\rightarrow$	$Ba\,(NO_3)_2(aq) + 2\,AgCl(s)$
Initial:	0.750 mol		0.700 mol		0.000 mol
Change:	-0.350 mol		-0.700 mol		$+0.350$ mol
Final:	0.400 mol		0.000 mol		0.350 mol

The AgCl is not in solution.

213. The Ba^{2+} ion and the NO_3^- ion do not react (they are spectator ions), so there are 0.750 mol of Ba^{2+} ion and 0.70 mol of NO_3^- ion in the final solution. The numbers of moles of silver ion and chloride ion are calculated using the net ionic equation:

	Cl^-	$+$	Ag^+	$\rightarrow$	$AgCl(s)$
Initial:	1.50		0.70		
Change:	-0.70		-0.70		
Final:	0.80		0.00		

Using the net ionic equation makes this calculation simpler. (Compare to Question 212.) Note that 0.400 mol $BaCl_2$ and 0.350 mol $Ba(NO_3)_2$ in the prior question contain a total of 0.750 mol of Ba^{2+} in solution.

Chapter 12: Measures of Concentration

214. $0.200 \text{ mol} \left(\dfrac{1\,L}{1.71 \text{ mol}} \right) = 0.117\,L = 117\text{ mL}$

215. $(84.0 \text{ g NaOH}) \left(\dfrac{1 \text{ mol NaOH}}{40.0 \text{ g NaOH}} \right) \left(\dfrac{1\,L}{3.00 \text{ mol NaOH}} \right) = 0.700\,L$

216. $(24.0 \text{ g NaOH})\left(\dfrac{1 \text{ mol NaOH}}{40.0 \text{ g NaOH}}\right) = 0.600 \text{ mol NaOH}$

$\dfrac{0.600 \text{ mol}}{0.300 \text{ L}} = 2.00 \text{ M}$

217. $(0.400 \text{ mol})\left(\dfrac{1 \text{ L}}{2.50 \text{ mol}}\right) = 0.160 \text{ L}$

218. $(200 \text{ mL HCl})\left(\dfrac{0.650 \text{ mmol HCl}}{\text{mL}}\right) = 130 \text{ mmol HCl}$

$130 \text{ mmol}\left(\dfrac{1 \text{ mL}}{0.200 \text{ mmol}}\right) = 650 \text{ mL}$

Approximately $(650 \text{ mL}) - (200 \text{ mL}) = 450 \text{ mL}$ must be added.

219. $(2.0 \text{ L})\left(\dfrac{0.40 \text{ mol Na}^+}{\text{L}}\right)\left(\dfrac{1 \text{ mol Na}_2\text{SO}_4}{2 \text{ mol Na}^+}\right)\left(\dfrac{1 \text{ L}}{0.30 \text{ mol Na}_2\text{SO}_4}\right) = 1.3 \text{ L}$

220. $(0.20 \text{ mol}) + (2.0 \text{ L})\left(\dfrac{1.1 \text{ mol}}{\text{L}}\right) = 2.4 \text{ mol}$ $\qquad (2.4 \text{ mol})/(3.0 \text{ L}) = 0.80 \text{ M}$

221. A 1.000 M solution contains 1.000 mol solute in 1.000 L solution. $Pb(NO_3)_2$ is 331.2 g/mol; hence, 331.2 g $Pb(NO_3)_2$ is needed. A 1.000 M solution of $Pb(NO_3)_2$ is 1.000 M with respect to Pb^{2+} and 2.000 M with respect NO_3^-.

222. $100 \text{ g NaCl}\left(\dfrac{1 \text{ mol NaCl}}{58.5 \text{ g NaCl}}\right) = 1.71 \text{ mol NaCl}$ $\qquad \dfrac{1.71 \text{ mol}}{1.50 \text{ L}} = 1.14 \text{ M}$

223. **(1)** and **(4)**. **(1)** contains 5.0 mmol H^+ in 75 mL; **(2)** contains 5.0 mmol H^+ in 100 mL; **(3)** contains 10 mmol H^+ in 75 mL; **(4)** contains 5.0 mmol in 75 mL (assuming complete ionization of the H_2SO_4).

224. $(40.00 \text{ mL})(1.600 \text{ M HCl})$ yields 64.00 mmol H^+ and 64.00 mmol Cl^-

$(60.00 \text{ mL})(2.000 \text{ M NaOH})$ yields 120.0 mmol OH^- and 120.0 mmol Na^+

$$H^+ + OH^- \rightarrow H_2O$$

The 64.00 mmol of H^+ reacts with 64.00 mmol of OH^-, leaving 56.0 mmol of OH^- in the solution.

$$\dfrac{56.0 \text{ mmol OH}^-}{100.0 \text{ mL}} = 0.560 \text{ M OH}^-$$

$$\dfrac{64.00 \text{ mmol Cl}^-}{100.0 \text{ mL}} = 0.6400 \text{ M Cl}^-$$

$$\dfrac{120.0 \text{ mmol Na}^+}{100.0 \text{ mL}} = 1.200 \text{ M Na}^+$$

225. $2\,NaOH + H_2SO_4 \rightarrow Na_2SO_4 + 2\,H_2O$

$$(200.0 \text{ mL NaOH})\left(\frac{0.500 \text{ mmol NaOH}}{\text{mL}}\right)\left(\frac{1 \text{ mmol H}_2\text{SO}_4}{2 \text{ mmol NaOH}}\right)\left(\frac{1 \text{ mL H}_2\text{SO}_4}{0.300 \text{ mmol H}_2\text{SO}_4}\right)$$
$$= 167 \text{ mL H}_2\text{SO}_4$$

226. $(12.35 \text{ g NaHCO}_3)\left(\dfrac{1 \text{ mol NaHCO}_3}{84.01 \text{ g NaHCO}_3}\right)\left(\dfrac{1 \text{ mol HCl}}{1 \text{ mol NaHCO}_3}\right)\left(\dfrac{1000 \text{ mL}}{3.000 \text{ mol HCl}}\right)$
$$= 49.00 \text{ mL}$$

227. $10.0 \text{ g Zn}\left(\dfrac{1 \text{ mol Zn}}{65.4 \text{ g Zn}}\right)\left(\dfrac{1 \text{ mol H}_2\text{SO}_4}{1 \text{ mol Zn}}\right)\left(\dfrac{1 \text{ L}}{3.00 \text{ mol H}_2\text{SO}_4}\right) = 0.0510 \text{ L} = 51.0 \text{ mL}$

228. $NH_3 + H^+ \rightarrow NH_4^+$

(50.0 mL)(0.300 M) contains 15.0 mmol H^+ and 15.0 mmol Cl^-

(50.0 mL)(0.400 M) contains 20.0 mmol NH_3

The reaction yields 15.0 mmol of NH_4^+, leaving 5.0 mmol NH_3 unreacted. The 15.0 mmol of Cl^- is unchanged.

$$\frac{15.0 \text{ mmol NH}_4^+}{100 \text{ mL}} = 0.150 \text{ M NH}_4^+$$

$$\frac{15.0 \text{ mmol Cl}^-}{100 \text{ mL}} = 0.150 \text{ M Cl}^-$$

$$\frac{5.0 \text{ mmol NH}_3}{100 \text{ mL}} = 0.050 \text{ M NH}_3$$

229. $HCl + NaOH \rightarrow NaCl + H_2O$

$$(4.50 \times 10^{-3} \text{ L NaOH})\left(\frac{3.00 \text{ mol NaOH}}{\text{L NaOH}}\right)\left(\frac{1 \text{ mol HCl}}{1 \text{ mol NaOH}}\right) = 13.5 \times 10^{-3} \text{ mol HCl}$$

$$\frac{13.5 \times 10^{-3} \text{ mol HCl}}{2.50 \times 10^{-3} \text{ L HCl}} = 5.40 \text{ M HCl}$$

230. $BaCl_2 + Na_2SO_4 \rightarrow BaSO_4 + 2\,NaCl$

$$(1.756 \text{ g BaSO}_4)\left(\frac{1 \text{ mol BaSO}_4}{233 \text{ g BaSO}_4}\right)\left(\frac{1 \text{ mol Na}_2\text{SO}_4}{1 \text{ mol BaSO}_4}\right) = 7.54 \times 10^{-3} \text{ mol Na}_2\text{SO}_4$$

$$\frac{7.54 \times 10^{-3} \text{ mol Na}_2\text{SO}_4}{0.0400 \text{ L}} = 0.188 \text{ M Na}_2\text{SO}_4$$

231. This is a limiting quantities question.

$$(2.00 \text{ L})\left(\frac{1.50 \text{ mol CuSO}_4}{\text{L}}\right) = 3.00 \text{ mol CuSO}_4 \quad (40.0 \text{ g Al})\left(\frac{1 \text{ mol Al}}{27.0 \text{ g Al}}\right) = 1.48 \text{ mol Al}$$

$$2\,Al + 3\,CuSO_4 \rightarrow Al_2(SO_4)_3 + 3\,Cu$$

The copper is in excess (3 mol $CuSO_4$ would react with 2 mol Al if it were present).

$$(1.48 \text{ mol Al})\left(\frac{3 \text{ mol Cu}}{2 \text{ mol Al}}\right)\left(\frac{63.5 \text{ g Cu}}{\text{mol Cu}}\right) = 141 \text{ g Cu}$$

232. $(1.00 \text{ L})(0.10 \text{ M}) = 0.10 \text{ mol Cu}^{2+}$

$$(0.10 \text{ mol Cu}^{2+})\left(\frac{1 \text{ mol CuS}}{1 \text{ mol Cu}^{2+}}\right)\left(\frac{95.6 \text{ g CuS}}{\text{mol CuS}}\right) = 9.6 \text{ g CuS}$$

$$(0.10 \text{ mol Cu}^{2+})\left(\frac{2 \text{ mol H}^+}{1 \text{ mol Cu}^{2+}}\right) = 0.20 \text{ mol H}^+$$

$$\frac{0.20 \text{ mol H}^+}{1.00 \text{ L}} = 0.20 \text{ M H}^+ \quad \text{(Assuming no volume change of the liquid.)}$$

233. In a solution of $A + B + C + \cdots$,

$$x(A) = \frac{a \text{ mol A}}{(a \text{ mol A}) + (b \text{ mol B}) + (c \text{ mol C}) + \cdots}$$

Similarly, for the mole fractions of B and C, and so on, all the fractions having the same denominator. The total of all the mole fractions is thus

$$\frac{(a \text{ mol A}) + (b \text{ mol B}) + (c \text{ mol C}) + \cdots}{(a \text{ mol A}) + (b \text{ mol B}) + (c \text{ mol C}) + \cdots} = 1$$

234. (A) $(1.0 \text{ g H}_2)\left(\dfrac{1 \text{ mol H}_2}{2.0 \text{ g H}_2}\right) = 0.50 \text{ mol H}_2 \quad (8.0 \text{ g O}_2)\left(\dfrac{1 \text{ mol O}_2}{32.0 \text{ g O}_2}\right) = 0.25 \text{ mol O}_2$

$$(16.0 \text{ g CH}_4)\left(\frac{1 \text{ mol CH}_4}{16.0 \text{ g CH}_4}\right) = 1.00 \text{ mol CH}_4$$

$$x(H_2) = \frac{0.50 \text{ mol H}_2}{1.75 \text{ mol total}} = 0.29$$

(B) $C_3H_5(OH)_3$ is 92.0 g/mol; H_2O is 18.0 g/mol.

$$46.0 \text{ g glycerin}\left(\frac{1 \text{ mol}}{92.0 \text{ g}}\right) = 0.500 \text{ mol} \quad 36.0 \text{ g water}\left(\frac{1 \text{ mol}}{18.0 \text{ g}}\right) = 2.00 \text{ mol H}_2O$$

Total number of moles $= 0.500 \text{ mol} + 2.00 \text{ mol} = 2.50 \text{ mol}$

$$x(\text{glycerin}) = \frac{n(\text{glycerin})}{\text{total number of moles}} = \frac{0.500 \text{ mol}}{2.50 \text{ mol}} = 0.200$$

$$x(\text{water}) = \frac{n(\text{water})}{\text{total number of moles}} = \frac{2.00 \text{ mol}}{2.50 \text{ mol}} = 0.800$$

Check: Sum of mole fractions $= 0.200 + 0.800 = 1.000$

235. Per liter of solution: $(2.00 \text{ mol acid})(60.0 \text{ g/mol}) = 120 \text{ g acid}$

$$(1.000 \text{ L})(1.02 \text{ kg/L}) = 1.02 \text{ kg solution} = 1020 \text{ g solution}$$

$$(1020 \text{ g solution}) - (120 \text{ g acid}) = 900 \text{ g water}$$

$$(900 \text{ g H}_2\text{O})\left(\frac{1 \text{ mol H}_2\text{O}}{18.0 \text{ g H}_2\text{O}}\right) = 50.0 \text{ mol H}_2\text{O} \quad x(\text{acid}) = \frac{2.00 \text{ mol}}{(50.0 \text{ mol}) + (2.00 \text{ mol})} = 0.0385$$

236. $C_{12}H_{22}O_{11}$ is 342 g/mol.

$$m = \text{molality} = \frac{\text{moles of solute}}{\text{mass of solvent in kg}} = \frac{(20.0 \text{ g})/(342 \text{ g/mol})}{0.125 \text{ kg}} = 0.468 \text{ m}$$

237. (A) Assuming that water has a density of 1.00 g/mL,

$$(300 \text{ mL water})\left(\frac{1.00 \text{ g water}}{\text{mL water}}\right)\left(\frac{1 \text{ kg}}{10^3 \text{ g}}\right)\left(\frac{2.46 \text{ mol}}{\text{kg}}\right)\left(\frac{111 \text{ g CaCl}_2}{\text{mol}}\right) = 81.9 \text{ g CaCl}_2$$

(B) C_2H_5OH is 46.1 g/mol. Since the molality is 1.54 m, 1 kg water dissolves 1.54 mol alcohol. Then 2.50 kg water dissolves $(2.50 \text{ kg})(1.54 \text{ mol/kg}) = 3.85 \text{ mol}$ alcohol, and

$$\text{Mass of alcohol} = (3.85 \text{ mol})(46.1 \text{ g/mol}) = 177 \text{ g alcohol}$$

238. $(57.5 \text{ mL C}_2\text{H}_5\text{OH})\left(\frac{0.800 \text{ g}}{\text{mL}}\right)\left(\frac{1 \text{ mol C}_2\text{H}_5\text{OH}}{46.0 \text{ g C}_2\text{H}_5\text{OH}}\right) = 1.00 \text{ mol C}_2\text{H}_5\text{OH}$

$$(600 \text{ mL C}_6\text{H}_6)\left(\frac{0.900 \text{ g}}{\text{mL}}\right)\left(\frac{1 \text{ kg}}{1000 \text{ g}}\right) = 0.540 \text{ kg C}_6\text{H}_6$$

$$\frac{1.00 \text{ mol C}_2\text{H}_5\text{OH}}{0.540 \text{ kg C}_6\text{H}_6} = 1.85 \text{ mol/kg} = 1.85 \text{ m}$$

239. (A) Each milliliter of acid solution has a mass of 1.198 g and contains
$(0.270)(1.198 \text{ g}) = 0.323 \text{ g H}_2\text{SO}_4$
Since the H_2SO_4 is 98.1 g/mol,

$$0.323 \text{ g}\left(\frac{1 \text{ mol}}{98.1 \text{ g}}\right) = 3.29 \times 10^{-3} \text{ mol} \qquad \frac{3.29 \times 10^{-3} \text{ mol}}{1 \times 10^{-3} \text{ L}} = 3.29 \text{ M}$$

(B) From **(A)**, there is 323 g, or 3.29 mol, of solute per liter of solution. The mass of water in 1.00 L of solution is $1198 \text{ g} - 323 \text{ g} = 875 \text{ g H}_2\text{O}$.

$$\frac{3.29 \text{ mol H}_2\text{SO}_4}{0.875 \text{ kg H}_2\text{O}} = 3.76 \text{ mol/kg} = 3.76 \text{ m}$$

240. $120 \text{ g acid}\left(\dfrac{1 \text{ mol}}{60.0 \text{ g}}\right) = 2.00 \text{ mol}$ $\qquad$ $100 \text{ g } C_2H_5OH\left(\dfrac{1 \text{ mol}}{46.0 \text{ g}}\right) = 2.17 \text{ mol}$

(A) In water:

$$\text{Molality} = \frac{2.00 \text{ mol acid}}{0.100 \text{ kg water}} = 20.0 \text{ m} \qquad x(\text{acid}) = \frac{2.00 \text{ mol}}{(2.00 + 5.56) \text{ mol}} = 0.265$$

(B) In ethanol:

$$\text{Molality} = \frac{2.00 \text{ mol acid}}{0.100 \text{ kg ethanol}} = 20.0 \text{ m} \qquad x(\text{acid}) = \frac{2.00 \text{ mol}}{(2.00 + 2.17) \text{ mol}} = 0.480$$

241. Consider a sample containing 1.000 mol total. There is then 0.0450 mol solute and

$$0.955 \text{ mol } H_2O\left(\frac{18.0 \text{ g } H_2O}{1 \text{ mol } H_2O}\right) = 17.2 \text{ g } H_2O = 0.0172 \text{ kg } H_2O$$

$$\frac{0.0450 \text{ mol solute}}{0.0172 \text{ kg } H_2O} = 2.62 \text{ m solute}$$

242. There is 1.00 mol solute in 1.00 kg H_2O.

$$(1.00 \text{ kg})\left(\frac{10^3 \text{ g}}{\text{kg}}\right)\left(\frac{1 \text{ mol } H_2O}{18.0 \text{ g } H_2O}\right) = 55.6 \text{ mol } H_2O \qquad x = \frac{1.00}{1.00 + 55.6} = 0.0177$$

243. **(A)** $(100 \text{ g } H_2O)\left(\dfrac{1.10 \text{ mol}}{1000 \text{ g } H_2O}\right)\left(\dfrac{53.5 \text{ g}}{\text{mol}}\right) = 5.88 \text{ g}$

(B) $(100 \text{ g } H_2O)\left(\dfrac{25.0 \text{ g } NH_4Cl}{75.0 \text{ g } H_2O}\right) = 33.3 \text{ g}$

(C) $(100 \text{ g } H_2O)\left(\dfrac{1 \text{ mol } H_2O}{18.0 \text{ g } H_2O}\right) = 5.56 \text{ mol } H_2O$

$$(5.56 \text{ mol } H_2O)\left(\frac{0.150 \text{ mol } NH_4Cl}{0.850 \text{ mol } H_2O}\right)\left(\frac{53.5 \text{ g}}{\text{mol } NH_4Cl}\right) = 52.5 \text{ g } NH_4Cl$$

244. Consider 1.000 L of solution: Its mass is 1.251 kg = 1251 g. It contains

$$(1251 \text{ g solution})\left(\frac{35.0 \text{ g } HClO_4}{100.0 \text{ g solution}}\right)\left(\frac{1 \text{ mol } HClO_4}{100.5 \text{ g } HClO_4}\right) = 4.36 \text{ mol } HClO_4$$

$$(1.251 \text{ kg solution})\left(\frac{0.650 \text{ kg } H_2O}{1 \text{ kg solution}}\right) = 0.813 \text{ kg } H_2O$$

$$\frac{4.36 \text{ mol } HClO_4}{0.813 \text{ kg}} = 5.36 \text{ m} \qquad 4.36 \text{ mol/L} = 4.36 \text{ M}$$

245. In each 100.0 g of solution are

$$(10.0 \text{ g KCl})\left(\frac{1 \text{ mol}}{74.5 \text{ g}}\right) = 0.134 \text{ mol KCl} \quad \text{and} \quad (90.0 \text{ g H}_2\text{O})\left(\frac{1 \text{ mol}}{18.0 \text{ g}}\right) = 5.00 \text{ mol H}_2\text{O}$$

$$\text{Mole fraction} = \frac{0.134 \text{ mol}}{(0.134 \text{ mol}) + (5.00 \text{ mol})} = 0.0261$$

$$(100.0 \text{ g solution})\left(\frac{1 \text{ mL}}{1.06 \text{ g}}\right) = 94.3 \text{ mL} = 0.0943 \text{ L}$$

$$\text{Molarity} = \frac{0.134 \text{ mol KCl}}{0.0943 \text{ L}} = 1.42 \text{ M} \qquad \text{Molality} = \frac{0.134 \text{ mol KCl}}{0.0900 \text{ kg H}_2\text{O}} = 1.49 \text{ m}$$

Chapter 13: Gases

246.

	1	2
P	760 torr = 1.00 atm	1.25 atm
V	2.00 L	V_2

$$P_1V_1 = P_2V_2$$

$$V_2 = \frac{P_1V_1}{P_2} = \frac{(1.00 \text{ atm})(2.00 \text{ L})}{1.25 \text{ atm}} = 1.60 \text{ L}$$

247. Note that temperatures must be expressed in kelvins.

$$\frac{V_1}{T_1} = \frac{V_2}{T_2} \qquad V_2 = \frac{V_1 T_2}{T_1} = \frac{(4.00 \text{ L})(373 \text{ K})}{273 \text{ K}} = 5.47 \text{ L}$$

248.
$$\frac{P_1V_1}{T_1} = \frac{P_2V_2}{T_2} \qquad \frac{T_1}{P_1V_1} = \frac{T_2}{P_2V_2}$$

$$T_2 = \frac{T_1 P_2 V_2}{P_1 V_1} = \frac{(300 \text{ K})[(800/760) \text{ atm}](3.00 \text{ L})}{(1.00 \text{ atm})(2.00 \text{ L})} = 474 \text{ K}$$

249.
$$\frac{P_1V_1}{T_1} = \frac{P_2V_2}{T_2} \qquad V_2 = \frac{P_1V_1T_2}{T_1P_2} = \frac{(785 \text{ torr})(350 \text{ mL})(273 \text{ K})}{(353 \text{ K})(760 \text{ torr})} = 280 \text{ mL}$$

250. $6.00 \text{ L}\left(\frac{1 \text{ mol (STP)}}{22.4 \text{ L}}\right)\left(\frac{17.0 \text{ g}}{\text{mol}}\right) = 4.55 \text{ g}$

251. $V = \dfrac{nRT}{P} = \dfrac{(0.3000 \text{ mol})(0.0821 \text{ L} \cdot \text{atm/mol} \cdot \text{K})(333 \text{ K})}{0.821 \text{ atm}} = 9.99 \text{ L}$

252. $(8.40 \text{ g N}_2)\left(\dfrac{1 \text{ mol N}_2}{28.0 \text{ g N}_2}\right) = 0.300 \text{ mol N}_2$

$V = \dfrac{nRT}{P} = \dfrac{(0.300 \text{ mol})(0.0821 \text{ L} \cdot \text{atm/mol} \cdot \text{K})(373 \text{ K})}{(800/760) \text{ atm}} = 8.73 \text{ L}$

253. Do not use the ideal gas law for this question because water at 4°C and 1.00 atm is not a gas.

$$(18.0 \text{ g})\left(\dfrac{1.00 \text{ mL}}{1.00 \text{ g}}\right) = 18.0 \text{ mL}$$

254. The volume is found from the initial conditions:

$$n = (370 \text{ g})\left(\dfrac{1 \text{ mol}}{32.0 \text{ g}}\right) = 11.6 \text{ mol}$$

$$V = \dfrac{nRT}{P} = \dfrac{(11.6 \text{ mol})(0.0821 \text{ L} \cdot \text{atm/mol} \cdot \text{K})(298 \text{ K})}{3.00 \text{ atm}} = 94.6 \text{ L}$$

The final number of moles is $n = \dfrac{PV}{RT} = \dfrac{(1.00 \text{ atm})(94.6 \text{ L})}{(0.0821 \text{ L} \cdot \text{atm/mol} \cdot \text{K})(348 \text{ K})} = 3.31 \text{ mol}$

$(3.31 \text{ mol})\left(\dfrac{32.0 \text{ g}}{\text{mol}}\right) = 106 \text{ g}$ $(370 \text{ g initial}) - (106 \text{ g final}) = 264 \text{ g escaped}$

255. In each liter, the number of moles of molecules is given by

$$n = \dfrac{PV}{RT} = \dfrac{(1.00 \text{ atm})(1.00 \text{ L})}{(0.0821 \text{ L} \cdot \text{atm/mol} \cdot \text{K})(300 \text{ K})} = 0.0406 \text{ mol}$$

$$\dfrac{3.17 \text{ g}}{0.0406 \text{ mol}} = 78.1 \text{ g/mol}$$

The molar mass of HF is 20 g/mol. The large apparent molar mass from the gas density data means that the gas is appreciably associated even in the gas phase, presumably by hydrogen bonding. On average, about 4 HF molecules are hydrogen bonded together under these conditions.

256. $(1.00 \text{ g H}_2)\left(\dfrac{1 \text{ mol H}_2}{2.00 \text{ g H}_2}\right) = 0.500 \text{ mol H}_2$

$(88.0 \text{ g CO}_2)\left(\dfrac{1 \text{ mol CO}_2}{44.0 \text{ g CO}_2}\right) = 2.00 \text{ mol CO}_2$

$$P(\text{H}_2) = \dfrac{n(\text{H}_2)RT}{V} = \dfrac{(0.500 \text{ mol})(0.0821 \text{ L} \cdot \text{atm/mol} \cdot \text{K})(300 \text{ K})}{30.0 \text{ L}} = 0.410 \text{ atm}$$

$$P(\text{CO}_2) = \dfrac{(2.00 \text{ mol})(0.0821 \text{ L} \cdot \text{atm/mol} \cdot \text{K})(300 \text{ K})}{30.0 \text{ L}} = 1.64 \text{ atm}$$

$P(\text{total}) = (0.410 \text{ atm}) + (1.64 \text{ atm}) = 2.05 \text{ atm}$

257. The final total volume was 500 mL.

$$Oxygen: \quad P_{final} = P_{initial}\left(\frac{V_{initial}}{V_{final}}\right) = (220 \text{ torr})\left(\frac{200 \text{ mL}}{500 \text{ mL}}\right) = 88.0 \text{ torr}$$

$$Nitrogen: \quad P_f = P_i\left(\frac{V_i}{V_f}\right) = (100 \text{ torr})\left(\frac{300 \text{ mL}}{500 \text{ mL}}\right) = 60.0 \text{ torr}$$

$$\text{Total pressure} = (88.0 \text{ torr}) + (60.0 \text{ torr}) = 148.0 \text{ torr}$$

258. $P(O_2) = (787 \text{ torr}) - (27 \text{ torr}) = 760 \text{ torr} = 1.00 \text{ atm}$

$$n(O_2) = \frac{PV}{RT} = \frac{(1.00 \text{ atm})(3.00 \text{ L})}{(0.0821 \text{ L} \cdot \text{atm/mol} \cdot \text{K})(300 \text{ K})} = 0.122 \text{ mol}$$

$$V = \frac{nRT}{P} = \frac{(0.122 \text{ mol})(0.0821 \text{ L} \cdot \text{atm/mol} \cdot \text{K})(273 \text{ K})}{1.00 \text{ atm}} = 2.73 \text{ L}$$

259. The gas collected is a mixture of oxygen and water vapor.

$$\text{Pressure of dry oxygen} = (\text{total pressure}) - (\text{vapor pressure of water})$$
$$= (800 \text{ torr}) - (21.1 \text{ torr}) = 779 \text{ torr}$$

Thus, for the dry oxygen, $V_1 = 100 \text{ mL}$, $T_1 = 23 + 273 = 296 \text{ K}$, $P_1 = 779 \text{ torr}$. Converting to STP,

$$V_2 = V_1\left(\frac{T_2}{T_1}\right)\left(\frac{P_1}{P_2}\right) = (100 \text{ mL})\left(\frac{273 \text{ K}}{296 \text{ K}}\right)\left(\frac{779 \text{ torr}}{760 \text{ torr}}\right) = 94.5 \text{ mL}$$

260. $P(O_2) = P(\text{total}) - P(\text{water}) = (740 \text{ torr}) - (24 \text{ torr}) = 716 \text{ torr}$

$$n = \frac{PV}{RT} = \frac{(\frac{716}{760} \text{ atm})(10.5 \text{ L})}{(0.0821 \text{ L} \cdot \text{atm/mol} \cdot \text{K})(298 \text{ K})} = 0.404 \text{ mol} \quad (0.404 \text{ mol})\left(\frac{32.0 \text{ g}}{\text{mol}}\right) = 12.9 \text{ g}$$

261. $n = \frac{PV}{RT} = \frac{(1.00 \text{ atm})(0.560 \text{ L})}{(0.0821 \text{ L} \cdot \text{atm/K} \cdot \text{mol})(273 \text{ K})} = 0.0250 \text{ mol}$

$$\frac{1.80 \text{ g}}{0.0250 \text{ mol}} = 72.0 \text{ g/mol}$$

262. $Ca + 2HCl \rightarrow CaCl_2 + H_2$

$$(12.2 \text{ g Ca})\left(\frac{1 \text{ mol Ca}}{40.08 \text{ g Ca}}\right)\left(\frac{1 \text{ mol H}_2}{1 \text{ mol Ca}}\right) = 0.304 \text{ mol H}_2$$

$$V = \frac{nRT}{P} = \frac{(0.304 \text{ mol})(0.0821 \text{ L} \cdot \text{atm/mol} \cdot \text{K})(291 \text{ K})}{1.00 \text{ atm}} = 7.26 \text{ L}$$

263. $2 \text{KClO}_3(s) \rightarrow 2 \text{KCl} + 3 \text{O}_2(g)$

$$110 \text{ g KClO}_3 \left(\frac{1 \text{ mol KClO}_3}{122.6 \text{ g KClO}_3} \right) \left(\frac{3 \text{ mol O}_2}{2 \text{ mol KClO}_3} \right) = 1.35 \text{ mol O}_2$$

$$V = \frac{nRT}{P} = \frac{(1.35 \text{ mol})(0.0821 \text{ L} \cdot \text{atm/mol} \cdot \text{K})(291 \text{ K})}{(750/760) \text{ atm}} = 32.7 \text{ L}$$

264. $2 \text{KClO}_3 \rightarrow 2 \text{KCl} + 3 \text{O}_2$

$P(\text{O}_2) = (760 \text{ torr}) - (19.8 \text{ torr}) = 740 \text{ torr}$

$$n(\text{O}_2) = \frac{PV}{RT} = \frac{[(740/760) \text{ atm}](1.80 \text{ L})}{(0.0821 \text{ L} \cdot \text{atm/mol} \cdot \text{K})(295 \text{ K})} = 0.0724 \text{ mol O}_2$$

$2 \text{KClO}_3 \rightarrow 2 \text{KCl} + 3 \text{O}_2$

$$(0.0724 \text{ mol O}_2) \left(\frac{2 \text{ mol KClO}_3}{3 \text{ mol O}_2} \right) \left(\frac{122.6 \text{ g KClO}_3}{\text{mol KClO}_3} \right) = 5.92 \text{ g KClO}_3$$

265. (A) $(450 \text{ g KClO}_3) \left(\frac{1 \text{ mol KClO}_3}{122.6 \text{ g KClO}_3} \right) \left(\frac{3 \text{ mol O}_2}{2 \text{ mol KClO}_3} \right) = 5.51 \text{ mol O}_2$

$$V = \frac{nRT}{P} = \frac{(5.51 \text{ mol})(0.0821 \text{ L} \cdot \text{atm/mol} \cdot \text{K})(293 \text{ K})}{0.996 \text{ atm}} = 133 \text{ L}$$

(B) $17.5 \text{ torr} \left(\frac{1 \text{ atm}}{760 \text{ torr}} \right) = 0.0230 \text{ atm}$; hence, the partial pressure of O_2 is

$0.996 \text{ atm} - 0.0230 \text{ atm} = 0.973 \text{ atm}$

$$V = \frac{nRT}{P} = \frac{(5.51 \text{ mol})(0.0821 \text{ L} \cdot \text{atm/mol} \cdot \text{K})(293 \text{ K})}{0.973 \text{ atm}} = 136 \text{ L}$$

266. Before the reaction there are 1.040 mol of NO and

$$(20.0 \text{ g O}_2) \left(\frac{1 \text{ mol O}_2}{32.0 \text{ g O}_2} \right) = 0.625 \text{ mol O}_2.$$

$n(\text{total}) = 1.665 \text{ mol} \quad P = \frac{nRT}{V} = \frac{(1.665 \text{ mol})(0.0821 \text{ L} \cdot \text{atm/mol} \cdot \text{K})(300 \text{ K})}{20.0 \text{ L}} = 2.05 \text{ atm}$

After the reaction, 1.040 mol of NO_2 would be present; 0.105 mol of oxygen would remain unreacted. The total number of moles of gas is 1.145 mol. The pressure is

$$P = \frac{nRT}{V} = \frac{(1.145 \text{ mol})(0.0821 \text{ L} \cdot \text{atm/mol} \cdot \text{K})(300 \text{ K})}{20.0 \text{ L}} = 1.41 \text{ atm}$$

There is a 0.64 atm pressure reduction, which results from the reaction of a greater number of moles of gas to yield a lesser number of moles of gas.

267. $2CO(g) + O_2(g) \rightarrow 2CO_2(g)$ $(112 \text{ g})\left(\dfrac{1 \text{ mol}}{28.0 \text{ g}}\right) = 4.00 \text{ mol CO present}$

$(48.0 \text{ g } O_2)\left(\dfrac{1 \text{ mol}}{32.0 \text{ g}}\right) = 1.50 \text{ mol } O_2 \text{ present}$

O_2 is in limiting quantity; CO is in excess.

$$(1.50 \text{ mol } O_2)\left(\dfrac{2 \text{ mol } CO_2}{\text{mol } O_2}\right) = 3.00 \text{ mol } CO_2$$

$$V = \frac{nRT}{P} = \frac{(3.00 \text{ mol})(0.0821 \text{ L} \cdot \text{atm/mol} \cdot \text{K})(273 \text{ K})}{1.00 \text{ atm}} = 67.2 \text{ L}$$

268. **(A)** Only the oxygen is gaseous; the carbon exerts no pressure.

$$P = \frac{nRT}{V} = \frac{(0.123 \text{ mol})(0.0821 \text{ L} \cdot \text{atm/mol} \cdot \text{K})(298 \text{ K})}{3.00 \text{ L}} = 1.00 \text{ atm}$$

(B) $2C + O_2 \rightarrow 2CO$

Two moles of CO is produced for every mol O_2 used up, yielding 0.246 mol CO. The final pressure is

$$P = \frac{(0.246 \text{ mol})(0.0821 \text{ L} \cdot \text{atm/mol} \cdot \text{K})(298 \text{ K})}{3.00 \text{ L}} = 2.01 \text{ atm}$$

269. **(A)** $P = \dfrac{nRT}{V} = \dfrac{(0.60 \text{ mol})(0.0821 \text{ L} \cdot \text{atm/mol} \cdot \text{K})(298 \text{ K})}{3.00 \text{ L}} = 4.9 \text{ atm}$

(B) $P = \dfrac{nRT}{V - nb} - \dfrac{n^2 a}{V^2}$

$= \dfrac{(0.60 \text{ mol})(0.0821 \text{ L} \cdot \text{atm/mol} \cdot \text{K})(298 \text{ K})}{(3.00 \text{ L}) - (0.60 \text{ mol})(0.0371 \text{ L/mol})} - \dfrac{(0.60 \text{ mol})^2(4.17 \text{ L}^2 \cdot \text{atm/mol}^2)}{(3.00 \text{ L})^2}$

$= 4.8 \text{ atm}$

270. $$\left(P + \frac{n^2 a}{V^2}\right)(V - nb) = nRT$$

From Table 13.1, $a = 3.59 \text{ L}^2 \cdot \text{atm/mol}$ and $b = 0.0427 \text{ L/mol}$.

$$\left(P + \frac{100 \times 3.59}{4.00} \text{ atm}\right)\left[(2.00 \text{ L}) - (10.0 \text{ mol})(0.0427 \text{ L/mol})\right] = 10.0(0.0821)(320) \text{ L} \cdot \text{atm}$$

$$P = 77 \text{ atm}$$

From the ideal gas equation: $PV = nRT$ $(2.00 \text{ L})P = (10.0)(0.0821)(320) \text{ L} \cdot \text{atm}$

$$P = 131 \text{ atm}$$

Since the total pressure is so high, one would not expect the gas to behave ideally. The van der Waals equation gives a result much closer to the experimental value.

271. $\left(P + \dfrac{n^2 a}{V^2}\right)(V - nb) = nRT$ compared to $PV = nRT$

(A) At high pressure, the volume is low. Hence, the $n^2 a/V^2$ term is more important than at low pressure and high volume. (The square term makes the change in pressure more important for $n^2 a/V^2$ than for P.) The constant nb term is also more important when subtracted from the smaller V. (B) At low temperatures and a given pressure, V must be low, and the effect is as in part (A).

272. The constant a is related to intermolecular forces; those for NH_3 are much higher because of hydrogen bonding. The constant b is related to molecular volume. The hydrogen atoms of NH_3 take up practically no volume, and the ammonia molecule is little over half the volume of that of N_2.

273. (A) The facts that the molecules of each gas are in constant random motion and exert no forces on each other are necessary to have pressures independent of the presence of other gases. (B) The average translational kinetic energy is directly proportional to absolute temperature.

$$m_1 u_1^2 = m_2 u_2^2 \qquad\qquad \frac{u_2}{u_1} = \sqrt{\frac{m_1}{m_2}} = \sqrt{\frac{M_1}{M_2}}$$

274. $R = N_A k = (6.02 \times 10^{23})k$

275. The constant a reflects the intermolecular attractions in real gases. The kinetic molecular theory postulates no intermolecular attractions, which is a good approximation in many cases. The constant b reflects the actual volume of the molecules themselves. The kinetic molecular theory postulates that the volume of the molecules is zero (or negligible), also sometimes a good approximation.

276. NH_3, which has the highest intermolecular forces—hydrogen bonding—would take the greatest quantity of energy to overcome those forces. Hence, it would take the most heat to restore the original temperature.

277. The gas does work pushing back the atmosphere and overcoming intermolecular attractions. The energy comes from the kinetic energy of the molecules, lowering the average kinetic energy—the temperature.

278. Quadrupling the absolute temperature causes a quadrupling in the average kinetic energy of the molecules. This increase in average KE causes an increase of the root mean square speed by a factor of 2:

$$\overline{KE}_1 = \tfrac{1}{2} m u_1^2 \qquad\qquad \overline{KE}_2 = \tfrac{1}{2} m u_2^2 = 4\overline{KE}_1 = \tfrac{1}{2} m (2u_1)^2$$

Thus, the molecules are traveling twice as fast at the higher temperature; as a consequence, they hit the walls twice as often. In addition, because they are traveling twice as fast, their momentum is doubled, and their change in momentum at the wall also doubles—they hit the wall twice as hard each time. The combined effect of hitting the wall twice as often with twice the change in momentum each time they hit causes a fourfold increase in pressure, in agreement with Charles' law.

279. Consider the following simplified analysis. (Rigorous analysis provides the same result.) The volume of the box will be one-eighth the original volume: $(l/2)^3 = V/8$. With no change in temperature, the molecules will retain the original average kinetic energy and thus the original mean square speed: $\overline{KE} = \frac{3}{2}kT = \frac{1}{2}mu^2$. The molecules will strike the wall twice as often, since the distance in any given direction is only half as great. Also, the area of the wall will be reduced to one-fourth of the original area, and the pressure (force per unit area) will be increased on that account by a factor of 4. Thus, the pressure due to collisions on the wall will be increased by a factor of 2 because the molecules hit the wall more often and by a factor of 4 because the wall is smaller, for a total increase of a factor of 8. This result is exactly that predicted by Boyle's law.

280. $\dfrac{r(H_2)}{r(O_2)} = \sqrt{\dfrac{32.0}{2.0}} = 4.0$ Note the *inverse* relationship. H_2 effuses four times as fast as O_2 under these conditions.

281. $\dfrac{r(He)}{r(CO)} = \sqrt{\dfrac{\mathcal{M}(CO)}{\mathcal{M}(He)}} = \sqrt{\dfrac{28.0}{4.00}} = \sqrt{7.00} = 2.65$ Helium escapes at a rate 2.65 times as fast as CO.

$$r(CO) = \frac{r(He)}{2.65} = \frac{6.4 \text{ mmol He/h}}{2.65} = 2.4 \text{ mmol CO/h}$$

$$(10 \text{ mmol CO})\left(\frac{1 \text{ h}}{2.4 \text{ mmol CO}}\right) = 4.2 \text{ h}$$

282. The rate of pressure drop is directly proportional to the rate of effusion.

$$\frac{r(Ne)}{r(NH_3)} = \sqrt{\frac{\mathcal{M}(NH_3)}{\mathcal{M}(Ne)}} = \sqrt{\frac{17.0}{20.2}} = 0.917$$

$$r(NH_3) = r(Ne)/0.917 = \frac{0.30 \text{ torr/s}}{0.917} = 0.33 \text{ torr/s}$$

Chapter 14: Solids and Liquids

283. A cubic unit cell must have the six faces the same. The highest symmetry possible for an end-centered unit cell is tetragonal.

284. Uncharged atoms and molecules pack more efficiently in closest-packed structures.

285. The two structures have the same coordination number, hence the same packing fraction.

286. The cesium chloride unit cell is a simple cubic unit cell. (The center of the unit cell is occupied by an ion of opposite charge from the ions at the cell corners and thus is not a lattice point.)

287. **(A)** One (One Cs atom at the center, and one $[8 \times \frac{1}{8}]$ Cl atom at the corners.)

(B) Eight

288. $(1 \text{ CsCl unit}) \left(\dfrac{168.4 \text{ g}}{6.022 \times 10^{23} \text{units}} \right) \left(\dfrac{1 \text{ cm}^3}{3.988 \text{ g}} \right) = 7.012 \times 10^{-23} \text{ cm}^3$

289. X-rays

290. All the electrons in diamond form single bonds, located between two carbon atoms, and thus they are not free to move. One-quarter of the electrons in graphite form delocalized double bonds and move freely under the influence of an electrical potential.

291. The big difference in melting points suggests a difference in type of crystal binding. The intermolecular forces in solid CO_2 must be very low to be overcome in a low-temperature sublimation. CO_2 is actually a molecular lattice held together only by the weak van der Waals forces between discrete CO_2 molecules. SiO_2, on the other hand, is a covalent lattice, with a three-dimensional network of bonds; each silicon atom is bonded tetrahedrally to 4 oxygen atoms, and each oxygen atom is bonded to 2 silicon atoms.

292. **(A)** As heat is added, some ice melts completely; the rest stays completely solid. **(B)** Chocolate softens throughout, gradually. **(C)** The chocolate type. **(D)** Examples include glass, wax, butter, plastics, molasses, rubber.

293. Crystalline solids have a regular submicroscopic arrangement of their particles (somewhat like soldiers on parade), while amorphous solids are much less regular in their submicroscopic arrangements (somewhat like small groups of soldiers marching among regular commuters in a crowded railroad station).

294. **(A)** The triple point is the point at which solid, liquid, and vapor of a single substance are in equilibrium with each other. The freezing point is the point at which solid and liquid are in equilibrium under 1 atm total pressure. (Some other substance must be present to achieve 1 atm pressure.) **(B)** The freezing point is higher for most substances, which have a positive slope of the liquid–solid equilibrium line in the phase diagram. (Water is a notable exception.)

295. The uncovered water will cool faster, since some of it can evaporate that way. Evaporation cools, since the most energetic molecules are the ones that are lost, and they use energy to push back the atmosphere.

296. The surface tension pulls the water into the capillary. In a fine capillary, the surface tension is great enough to overcome the attraction of gravity on the water.

297. The drops would be spherical—the shape with the smallest surface area per unit volume, in response to surface tension.

298. (A) Line $\overline{DB}$ represents the solid–gas equilibrium.

(B) When a sample at point E is heated, melting and later vaporization occur.

(C) The freezing point is the temperature at which line $\overline{BC}$ crosses the $P = 1$ atm line.

299. $H_2O(l) \rightleftharpoons H_2O(s)$

As the pressure increases at constant temperature, this equilibrium shifts to the liquid phase—the more dense phase. Thus a movement straight up on the phase diagram, increasing the pressure at constant temperature, causes melting as one crosses the solid–liquid equilibrium line. The slope is correct.

Chapter 15: Oxidation and Reduction

300. (A) The oxidation number is −2, which is equal to the charge on the monatomic ion. (B) In any neutral species, the oxidation numbers must total zero. With 2 hydrogen at +1, 4 oxygen at −2, sulfur at x,

$$2(+1) + x + 4(-2) = 0 \qquad x = +6 \qquad \text{The oxidation number of sulfur is +6.}$$

(C) In $S_2O_3{}^{2-}$, the oxidation numbers must total −2, the charge on the ion. With 2 sulfur at x and 3 oxygen at −2,

$$2x + 3(-2) = -2 \qquad x = +2 \qquad \text{The oxidation number of sulfur is +2.}$$

(D) From just the usual rules, any assignment of oxidation numbers that gives a net charge of zero could be made. However, noting that sulfur is in the same periodic group with oxygen, and by analogy with CO_2, the sulfur is usually assigned an oxidation number of −2. Then, two sulfur at −2 = −4, carbon at +4 = +4, and the net charge = 0. (E) The oxidation number of any free element is zero. (F) The oxidation numbers of alkali and alkaline earth metals in their compounds are +1 and +2, respectively. The oxidation number of sulfur can be established on this basis:

$$2(+1) + 4x + 6(-2) = 0 \qquad x = 2.5$$

(G) The oxidation number of chlorine is −1, as in all halogen compounds with elements other than oxygen or a heavier halogen.

$$2x + 2(-1) = 0 \qquad x = +1$$

Therefore, the oxidation number of sulfur is +1.

301. Hydrogen has a +1 oxidation state in all its compounds except those with group IA, IIA, or IIIA metals. It has 0 oxidation state as an element. (A) +1 (B) +1 (C) −1 (D) 0 (E) −1

302. The wetting agent is your face; the drying agent is the towel. (The drying agent gets wet during the process, since the water must go into it; the wetting agent gets dry.) In a redox reaction, the oxidizing agent is reduced; the reducing agent is oxidized. (The electrons are transferred in a manner similar to the water in the drying analogy.)

303. **(A)**, **(C)**, **(D)**, **(E)** **(B)** is a physical change; **(F)** is the change of an element in one form [oxidation state = 0] to another form [oxidation state = 0]; **(G)** is the reaction of an acid with a base.)

304. Except for periodic groups IB and 0, the maximum oxidation state is the group number. The minimum oxidation state for the metals is zero; for the nonmetals it is equal to the group number minus 8.   **(A)** +5, 0   **(B)** +6, −2   **(C)** +7, 0   **(D)** +4, 0   **(E)** +3, 0

305. **(A)** CsI   **(B)** SF_6 The highest oxidation number of sulfur is +6; the negative oxidation number of fluorine is −1. Hence, the predicted compound is SF_6. (Others are SF_4 and SF_2.)

306. Equations **B**, **E**, and **G** represent disproportionations, in that in each reaction the oxidizing agent and the reducing agent are the same.

307. Fe^{2+} is oxidized to Fe^{3+} by the nitrate ion, which is the oxidizing agent. Nitrogen is reduced from nitrogen(V) in the nitrate to nitrogen(IV) in NO_2, by the reducing agent Fe^{2+}.

308. **(A)** iron(II) chloride and iron(III) chloride.   **(B)** uranium(III) chloride and uranium(VI) fluoride.   **(C)** copper(I) oxide and copper(II) oxide. Note especially that the Roman numerals represent the oxidation states, not the number of ions in the formula. **(D)** Nitrous acid and nitric acid are familiar examples of the nomenclature of acids, for which the Stock system, using oxidation numbers, is not commonly used.

309. The formulas are S^{2-}, SO_3^{2-}, SO_4^{2-}, $S_2O_3^{2-}$, respectively. The oxidation numbers of sulfur are −2, +4, +6, and +2, respectively. Since the charge on each species is 2−, it is apparent that the charge and oxidation number are not the same. The charge is the sum of the oxidation numbers of all the atoms in a species and refers to the species as a whole. Oxidation number refers to atoms individually. There is no direct relationship between the oxidation number of a single atom and the charge on the ion as a whole.

310. In H_2O_2 the oxygen exists in an intermediate oxidation state; it can be either oxidized or reduced, depending on the reagents used. It can disproportionate according to the equation $2H_2O_2 \rightarrow 2H_2O + O_2$.

311. **(A)** MnO_4^-, Ce^{4+}, $Cr_2O_7^{2-}$, CrO_4^{2-}, HNO_3, F_2 **(B)** I^-, Na, Fe^{2+} **(C)** Cl^-, Na^+, F^-

312. **(A)** The oxidizing agent is O_2; the reducing agent is K.   **(B)**, **(C)**, and **(D)** are not redox reactions.   **(E)** H_2O_2, which disproportionates, is both an oxidizing and a reducing agent.

313. **(A)** If an acid or base is one of the reactants or products, the solution is the same. If ammonia is present, the solution is basic; if ammonium ion is present, it is acidic. If metals that would form insoluble hydroxides are shown in their ionic form, the solution is acidic. Otherwise, use acidic. **(B)** (i) acidic (ii) basic (iii) acidic (iv) basic

314. *Ion-Electron Method* The reduction and oxidation half-reactions may be started:

$$MnO_4^- \to Mn^{2+} \qquad\qquad Cl^- \to Cl_2$$

The Mn is already balanced. The equation requires $4\,H_2O$ on the right to balance the oxygen atoms, then $8\,H^+$ on the left to balance the hydrogen. The oxidation balances by inspection.

$$8\,H^+ + MnO_4^- \to Mn^{2+} + 4\,H_2O \qquad\qquad 2\,Cl^- \to Cl_2$$

The net charge on the left of the reduction is $+8 - 1 = +7$, and on the right it is $+2$; therefore, 5 electrons must be added to the left. In the oxidation the net charge on the left is -2, and on the right it is 0; therefore, 2 electrons must be added to the right.

$$8\,H^+ + MnO_4^- + 5\,e^- \to Mn^{2+} + 4\,H_2O \qquad\qquad 2\,Cl^- \to Cl_2 + 2\,e^-$$

The multiplying factors are seen to be 2 and 5, respectively, to make $10\,e^-$ on each side.

$$16\,H^+ + 2\,MnO_4^- + 10\,e^- \to 2\,Mn^{2+} + 8\,H_2O$$
$$\underline{\hspace{2em} 10\,Cl^- \to 5\,Cl_2 + 10\,e^- \hspace{2em}}$$
$$16\,H^+ + 2\,MnO_4^- + 10\,Cl^- \to 2\,Mn^{2+} + 8\,H_2O + 5\,Cl_2$$

Since MnO_4^- was added as $KMnO_4$, $2\,MnO_4^-$ introduces $2\,K^+$ to the left side of the equation; and since K^+ does not react, the same number will appear on the right side. Since Cl^- was added as KCl, the $10\,Cl^-$ introduces $10\,K^+$ to each side of the equation. Since H^+ was added as H_2SO_4, $16\,H^+$ introduces $8\,SO_4^{2-}$ to each side of the equation. Then

$$16\,H^+ + 8\,SO_4^{2-} + 2\,K^+ + 2\,MnO_4^- + 10\,K^+ + 10\,Cl^- \to$$
$$2\,Mn^{2+} + 8\,H_2O + 5\,Cl_2 + 12\,K^+ + 8\,SO_4^{2-}$$

$$8\,H_2SO_4 + 2\,KMnO_4 + 10\,KCl \to 2\,MnSO_4 + 6\,K_2SO_4 + 5\,Cl_2 + 8\,H_2O$$

Oxidation-State Method Mn undergoes a change in oxidation state from $+7$ in MnO_4^- to $+2$ in Mn^{2+}. Cl undergoes a change in oxidation state from -1 in Cl^- to 0 in Cl_2. The electron balance diagrams are

$$Mn^{VII} + 5\,e^- \to Mn^{II}$$
$$2\,Cl^{(-1)} \to Cl_2^{(0)} + 2\,e^-$$

The multiplying factors are 2 and 5, just as in the previous method.

$$2\,Mn^{VII} + 10\,e^- \to 2\,Mn^{II}$$
$$10\,Cl^{(-1)} \to 5\,Cl_2^{(0)} + 10\,e^-$$

Hence, the coefficients of $KMnO_4$ and $MnSO_4$ are 2, of KCl is 10, of Cl_2 is 5.

$$2\,KMnO_4 + 10\,KCl \rightarrow 2\,MnSO_4 + 5\,Cl_2 \qquad\qquad \text{(incomplete)}$$

So far, no provision has been made for H_2O, H_2SO_4, and K_2SO_4. The 8 atoms of oxygen from $2\,KMnO_4$ form $8\,H_2O$. For $8\,H_2O$ we need 16 atoms of hydrogen, which can be furnished by $8\,H_2SO_4$. The 12 atoms of K ($10\,KCl + 2\,KMnO_4$) yield $6\,K_2SO_4$. Note that all the oxygen in the oxidizing agent is converted to water. The sulfate ion retains its identity throughout the reaction.

$$2\,KMnO_4 + 10\,KCl + 8\,H_2SO_4 \rightarrow 2\,MnSO_4 + 5\,Cl_2 + 6\,K_2SO_4 + 8\,H_2O$$

315. **(A)** The balancing of the reduction equation proceeds as follows.

$$Cr_2O_7^{2-} \rightarrow Cr^{3+}$$

$$Cr_2O_7^{2-} \rightarrow 2\,Cr^{3+}$$

$$Cr_2O_7^{2-} \rightarrow 2\,Cr^{3+} + 7\,H_2O$$

$$14\,H^+ + Cr_2O_7^{2-} \rightarrow 2\,Cr^{3+} + 7\,H_2O$$

$$14\,H^+ + Cr_2O_7^{2-} + 6\,e^- \rightarrow 2\,Cr^{3+} + 7\,H_2O \qquad \text{(balanced)}$$

The balancing of the oxidation equation proceeds as follows.

$$Cl^- \rightarrow Cl_2$$

$$2\,Cl^- \rightarrow Cl_2$$

$$2\,Cl^- \rightarrow Cl_2 + 2\,e^- \qquad \text{(balanced)}$$

Combining:

$$1 \times [14\,H^+ + Cr_2O_7^{2-} + 6\,e^- \rightarrow 2\,Cr^{3+} + 7\,H_2O]$$

$$3 \times [\qquad\qquad 2\,Cl^- \rightarrow Cl_2 + 2\,e^- \qquad\quad]$$

$$\overline{14\,H^+ + Cr_2O_7^{2-} + 6\,Cl^- \rightarrow 2\,Cr^{3+} + 7\,H_2O + 3\,Cl_2}$$

The $14\,H^+$ was added as $14\,HCl$, and 6 of the 14 chloride ions were oxidized. To each side of the equation, 8 more Cl^- can be added to represent those Cl^- that were not oxidized. Similarly, $2\,K^+$ may be added to each side to show that $Cr_2O_7^{2-}$ came from $K_2Cr_2O_7$.

$$14\,H^+ + 6\,Cl^- + 8\,Cl^- + 2\,K^+ + Cr_2O_7^{2-} \rightarrow 2\,Cr^{3+} + 2\,K^+ + 8\,Cl^- + 3\,Cl_2 + 7\,H_2O$$

$$14\,HCl + K_2Cr_2O_7 \rightarrow 2\,CrCl_3 + 2\,KCl + 3\,Cl_2 + 7\,H_2O$$

(B) $Fe^{2+} + H_2O_2 \rightarrow Fe^{3+} + H_2O$

The half-reactions:

$$Fe^{2+} \rightarrow Fe^{3+} + e^- \qquad\qquad\qquad 2\,H^+ + H_2O_2 \rightarrow 2\,H_2O$$

$$2\,Fe^{2+} \rightarrow 2\,Fe^{3+} + 2\,e^- \qquad\qquad 2\,H^+ + H_2O_2 + 2\,e^- \rightarrow 2\,H_2O$$

Combining half-reactions:

$$2\,Fe^{2+} + 2\,H^+ + H_2O_2 \rightarrow 2\,Fe^{3+} + 2\,H_2O$$

or $\quad\quad 2\,FeCl_2 + 2\,HCl + H_2O_2 \rightarrow 2\,FeCl_3 + 2\,H_2O$

(C) $Cu \rightarrow Cu^{2+} + 2\,e^-$ $\quad\quad\quad\quad\quad\quad\quad\quad NO_3^- \rightarrow NO$

$$NO_3^- \rightarrow NO + 2\,H_2O$$

$$4\,H^+ + NO_3^- \rightarrow NO + 2\,H_2O$$

$$3\,e^- + 4\,H^+ + NO_3^- \rightarrow NO + 2\,H_2O$$

$3\,Cu \rightarrow 3\,Cu^{2+} + 6\,e^- \quad\quad 6\,e^- + 8\,H^+ + 2\,NO_3^- \rightarrow 2\,NO + 4\,H_2O$

Combining half-reactions: $3\,Cu + 8\,H^+ + 2\,NO_3^- \rightarrow 2\,NO + 4\,H_2O + 3\,Cu^{2+}$

or $\quad\quad\quad\quad\quad\quad 3\,Cu + 8\,HNO_3 \rightarrow 2\,NO + 4\,H_2O + 3\,Cu(NO_3)_2$

316. (A) $Na_2C_2O_4 + KMnO_4 + H_2SO_4 \rightarrow K_2SO_4 + Na_2SO_4 + H_2O + MnSO_4 + CO_2$

$C_2O_4^{2-} \rightarrow CO_2 \quad\quad\quad\quad\quad\quad\quad\quad MnO_4^- \rightarrow Mn^{2+}$

$C_2O_4^{2-} \rightarrow 2\,CO_2 + 2\,e^- \quad\quad\quad\quad\quad MnO_4^- \rightarrow Mn^{2+} + 4\,H_2O$

$$8\,H^+ + MnO_4^- \rightarrow Mn^{2+} + 4\,H_2O$$

$$5\,e^- + 8\,H^+ + MnO_4^- \rightarrow Mn^{2+} + 4\,H_2O$$

$5\,C_2O_4^{2-} \rightarrow 10\,CO_2 + 10\,e^- \quad 10\,e^- + 16\,H^+ + 2\,MnO_4^- \rightarrow 2\,Mn^{2+} + 8\,H_2O$

$5\,C_2O_4^{2-} + 16\,H^+ + 2\,MnO_4^- \rightarrow 10\,CO_2 + 2\,Mn^{2+} + 8\,H_2O$

or $\quad 5\,Na_2C_2O_4 + 2\,KMnO_4 + 8\,H_2SO_4 \rightarrow 10\,CO_2 + 2\,MnSO_4 + 8\,H_2O + K_2SO_4 + 5\,Na_2SO_4$

(B) $I_2 + Na_2S_2O_3 \rightarrow Na_2S_4O_6 + NaI$

$\quad I_2 + S_2O_3^{2-} \rightarrow S_4O_6^{2-} + I^-$

$\quad 2\,e^- + I_2 \rightarrow 2\,I^- \quad\quad\quad\quad\quad\quad\quad\quad\quad 2\,S_2O_3^{2-} \rightarrow S_4O_6^{2-} + 2\,e^-$

$\quad\quad\quad\quad I_2 + 2\,S_2O_3^{2-} \rightarrow S_4O_6^{2-} + 2\,I^-$

or $\quad\quad\quad I_2 + 2\,Na_2S_2O_3 \rightarrow Na_2S_4O_6 + 2\,NaI$

(C) $NH_3 + O_2 \rightarrow NO + H_2O$

$\quad (-3 \rightarrow +2) = +5$

$\overline{NH_3 + O_2} \rightarrow \underline{NO + 2\,H_2O}$

$\quad\quad\quad 2(0 \rightarrow -2) = -4$

$\quad\quad 4\,NH_3 + 5\,O_2 \rightarrow 4\,NO + 6\,H_2O$

(D) $CuO + NH_3 \xrightarrow{\text{heat}} N_2 + H_2O + Cu$

$$(+2 \rightarrow 0) = -2$$

$$CuO + 2\underset{\underline{}}{N}H_3 \rightarrow N_2 + H_2O + Cu$$

$$2(-3 \rightarrow 0) = -6$$

$$3CuO + 2NH_3 \rightarrow N_2 + 3Cu$$

$$3CuO + 2NH_3 \rightarrow N_2 + 3Cu + 3H_2O$$

Chapter 16: Properties of Solutions

317. $P = xP°$ $\quad x(A) = \dfrac{P}{P°} = \dfrac{0.60 \text{ atm}}{0.80 \text{ atm}} = 0.75 \quad x(B) = 1 - x(A) = 1 - 0.75 = 0.25$

318. $20.0 \text{ g glucose} \left(\dfrac{1 \text{ mol glucose}}{180 \text{ g}} \right) = 0.111 \text{ mol glucose}$

$70.0 \text{ g } H_2O \left(\dfrac{1 \text{ mol } H_2O}{18.0 \text{ g } H_2O} \right) = 3.89 \text{ mol } H_2O$

$x(H_2O) = \dfrac{3.89}{0.111 + 3.89} = 0.972$

$P = P°x(H_2O) = (25.21 \text{ torr})(0.972) = 24.5 \text{ torr}$

319. Let z = number of moles of solute.

$$90.0 \text{ g } H_2O \left(\dfrac{1 \text{ mol } H_2O}{18.0 \text{ g } H_2O} \right) = 5.00 \text{ mol } H_2O$$

$$P = P°x$$

$$x = \dfrac{23.32 \text{ torr}}{23.76 \text{ torr}} = 0.981 = \dfrac{5.00}{5.00 + z}$$

$$(5.00 + z)(0.981) = 5.00$$

$$z = 0.10 \text{ mol}$$

$$\dfrac{5.40 \text{ g}}{0.10 \text{ mol}} = 54 \text{ g/mol}$$

320. Assuming ideal behavior,

$$P_b = x_b P_b° = 0.500 \, (119 \text{ torr}) = 59.5 \text{ torr}$$

$$P_t = x_t P_t° = 0.500 \, (37.0 \text{ torr}) = 18.5 \text{ torr}$$

$$P_{\text{tot}} = 78.0 \text{ torr}$$

$$\text{Mole fraction toluene in the vapor phase} = \dfrac{18.5 \text{ torr}}{78.0 \text{ torr}} = 0.237$$

321. 1.00 kg contains 0.100 kg CH_3OH (and 0.900 kg H_2O).

$$(100 \text{ g } CH_3OH)\left(\frac{1 \text{ mol}}{32.0 \text{ g}}\right) = 3.12 \text{ mol} \qquad \frac{3.12 \text{ mol}}{0.900 \text{ kg}} = 3.47 \text{ m}$$

$$\Delta t = K_f m = (1.86°C/m)(3.47 \text{ m}) = 6.45°C$$

$$\text{fp} = -6.45°C$$

322. (A) $\Delta t_b = K_b m = 0.284°C$

$$m = \frac{0.284°C}{2.11°C/m} = 0.135 \text{ m} = \frac{0.135 \text{ mol}}{\text{kg}}$$

$$\frac{10.6 \text{ g}}{0.740 \text{ kg ether}}\left(\frac{1 \text{ kg ether}}{0.135 \text{ mol}}\right) = 106 \text{ g/mol}$$

(B) $m = \Delta t_b/K_b = \Delta t_f/K_f = (0.50°C)/(0.512°C/m) = \Delta t_f/(1.86°C/m)$

$$\Delta t_f = 1.82°C \quad \text{from which } t = -1.82°C$$

323. $m = \dfrac{\Delta t}{K_f} = \dfrac{0.93°C}{1.86°C/m} = 0.50 \text{ m}$

$$\frac{36.0 \text{ g}}{1.20 \text{ kg } H_2O} = 30.0 \text{ g/kg } H_2O \qquad \frac{30.0 \text{ g}}{0.50 \text{ mol}} = 60 \text{ g/mol}$$

The empirical formula mass is 30 g/mol of empirical formula units; there must be 60 g/30 g = 2 empirical formula units per molecule. The formula is thus $(CH_2O)_2$ or $C_2H_4O_2$.

324. $\pi = \dfrac{n}{V}RT = \left(\dfrac{0.00100 \text{ mol}}{L}\right)\left(\dfrac{0.0821 \text{ L} \cdot \text{atm}}{\text{mol} \cdot K}\right)(273 \text{ K}) = 0.0224 \text{ atm} = 17.0 \text{ torr}$

Note that this value is easily measurable.

325. $\pi V = nRT = \dfrac{m}{\mathcal{M}_A}RT$

$$\mathcal{M}_A = \frac{mRT}{\pi V} = \frac{(10.0 \text{ g})(0.0821 \text{ L} \cdot \text{atm/mol} \cdot K)(300 \text{ K})}{[(10.0/760) \text{ atm}](1.00 \text{ L})} = 18,700 \text{ g/mol}$$

326. $\pi = \dfrac{nRT}{V} = \dfrac{(0.140 \text{ mol})(0.0821 \text{ L} \cdot \text{atm/mol} \cdot K)(300 \text{ K})}{0.250 \text{ L}} = 13.8 \text{ atm}$

327. (A) There are 2 mol of ions per mole of NaCl.   **(B)** There is interionic attraction, which reduces somewhat the independence of the ions and therefore lessens their effect on the freezing point. (The value of i is somewhat below 2.)

328. KCl has 2 ions per formula unit. $C_6H_{12}O_6$ has 1 molecule per formula unit. K_2SO_4 has 3 ions per formula unit. $Al_2(SO_4)_3$ has 5 ions per formula unit. NaCl has 2 ions per formula unit. $Al_2(SO_4)_3$ thus has the greatest freezing point depression and therefore the lowest freezing point.

329. (A) $m_{effective} = \dfrac{\Delta t_f}{K_f} = \dfrac{0.3433°C}{1.86°C/m} = 0.185$ m

$\Delta t_b = K_b m_{effective} = (0.512°C/m)(0.185$ m$) = 0.0947°C$

$t = (100.000 + 0.0947)°C = 100.095°C$

   (B) 0.00100 m NaClO$_3$ at such low concentrations is

0.00100 m Na$^+$ plus 0.00100 m ClO$_3^-$, a 0.00200 m solution:

$\Delta t_f = (1.86°C/m)(0.00200$ m$) = 0.00372°C \qquad t_f = -0.00372°C$

330. $\Delta t = iKm$

$$m = \left(\dfrac{100\text{ g KCl}}{\text{kg water}}\right)\left(\dfrac{1\text{ mol KCl}}{74.55\text{ g KCl}}\right) = 1.34\text{ m}$$

$$i = \dfrac{\Delta t}{Km} = \dfrac{4.5°C}{(1.86°C/m)(1.34\text{ m})} = 1.8$$

Chapter 17: Thermodynamics

331. Since heat is added to the system and work is done on the system, both q and w have positive values:

$$q = +678\text{ J} \qquad\qquad w = +294\text{ J}$$

$$\Delta E = q + w = 678\text{ J} + 294\text{ J} = 972\text{ J}$$

332. (A) Since the reaction is the reverse of the first given, the sign of ΔE is reversed, +10.0 kJ.

   (B) Since there is twice the number of moles of each reactant, the value of ΔE will be doubled also.

$$\Delta E = 20.0\text{ kJ}$$

   (C) Adding the two given equations, and canceling the substances that appear on both sides, yields the equation given in **(C)**. The values of ΔE are added also, yielding $\Delta E = +5.0$ kJ.

333. Heat = (mass)(specific heat)(temperature change) = (40.0 g)(0.895 J/g·°C) × (12.3°C)

$= 440$ J

334. Heat lost by alloy = heat absorbed by water

$(25.0$ g$)(c)[(100.0 - 27.2)°C] = (90.0$ g$)(4.184$ J/g·°C$)[(27.18 - 25.32)°C]$

$c = 0.385$ J/g·°C

335. Let t be the final temperature.

$$\text{Heat} = (\text{mass})(\text{specific heat})(\text{temperature change})$$

Heat loss by silver $= (100\text{ g})(0.236\text{ J/g} \cdot °\text{C})[(40.0 - t)°\text{C}]$ (Data from Table 17.1)

Heat gain by water $= (60.0\text{ g})(4.184\text{ J/g} \cdot °\text{C})[(t - 10.0)°\text{C}]$

Since the heat lost by the silver is gained by the water, the two products are equal:

$$(100\text{ g})(0.236\text{ J/g} \cdot °\text{C})(40.0 - t)°\text{C} = (60.0\text{ g})(4.184\text{ J/g} \cdot °\text{C})[(t - 10.0)°\text{C}]$$

$$t = 12.6°\text{C}$$

336. The added heat causes the following changes:

$$\text{solid} \xrightarrow{1} \text{solid} \xrightarrow{2} \text{liquid} \xrightarrow{3} \text{liquid}$$
$$-10°\text{C} \qquad 0°\text{C} \qquad 0°\text{C} \qquad +10°\text{C}$$

$$\Delta H_1 = m(c)\Delta t = 10.0\text{ g }(2.09\text{ J/g} \cdot °\text{C})(10.0°\text{C}) = 209\text{ J}$$
$$\Delta H_2 = m(\Delta H_{\text{fus}}) = 10.0\text{ g }(335\text{ J/g}) = 3350\text{ J}$$
$$\Delta H_3 = m(c)\Delta t = 10.0\text{ g }(4.184\text{ J/g} \cdot °\text{C})(10.0°\text{C}) = \underline{418\text{ J}}$$

$$\Delta H_{\text{total}} = 3977\text{ J}$$

337. (**1**) Consider the heat absorbed by the ice and by the water from it.

$$\Delta H(\text{fusion}) = (335\text{ J/g})(150\text{ g}) = 5.025 \times 10^4\text{ J}$$
$$\Delta H(\text{heating}) = mc\,\Delta t = (4.184\text{ J/g} \cdot °\text{C})(150\text{ g})[(t - 0)°\text{C}] = 628t$$

(**2**) Now consider the heat lost by the hot water.

$$\Delta H = mc\,\Delta t = (4.184\text{ J/g} \cdot °\text{C})(300\text{ g})[(t - 50.0)°\text{C}]$$

where $t < 50°\text{C}$

(**3**) The sum of the ΔH's must equal 0, since heat is assumed not to leak into or out of the total system treated in (1) and (2).

$$5.025 \times 10^4 + 628t + 1255(t - 50) = 0$$

$$t = 6.6°\text{C}$$

338. According to the law of Dulong and Petit, the molar heat capacity of Pt (or any other metallic element) is 25 J/mol $\cdot$ °C. The specific heat is therefore

$$\frac{25\text{ J}}{\text{mol} \cdot °\text{C}}\left(\frac{1\text{ mol}}{195\text{ g}}\right) = 0.13\text{ J/g} \cdot °\text{C}$$

339. From the law of Dulong and Petit, $C = 25$ J/mol·°C.

$$(1.6 \text{ mol})(25 \text{ J/mol·°C})(60.0 - t) = 100 \text{ g}(4.184 \text{ J/g·°C})(t - 20.0)$$

$$t = 23°C$$

340. Heat lost by metal = −heat gained by water

$$(40.0 \text{ g})c(37.0°C) = (100 \text{ g})(4.184 \text{ J/g·°C})(3.0°C)$$

$$c = 0.85 \text{ J/g·°C}$$

According to the law of Dulong and Petit, the metal has a molar heat capacity about 25 J/mol·°C. Hence the atomic mass is approximately

$$\frac{25 \text{ J}}{\text{mol·°C}}\left(\frac{1 \text{ g·°C}}{0.85 \text{ J}}\right) = 29 \text{ g/mol} = 29 \text{ u}$$

341. The number of moles of gaseous reactant (O_2) is $\frac{3}{2}$; the number of moles of gaseous products ($N_2 + CO_2$) is 2. Therefore, $\Delta n = 2 - \frac{3}{2} = \frac{1}{2}$.

$$\Delta H = \Delta E + \Delta(PV) = \Delta E + (\Delta n)RT$$

$$\Delta H = -742.7 \text{ kJ} + (0.500 \text{ mol})(8.314 \text{ J/mol·°C})(298°C) = -742.7 \text{ kJ} + 1240 \text{ J} = -741.5 \text{ kJ}$$

342. $(35.2 \text{ g CO}_2)\left(\dfrac{1 \text{ mol CO}_2}{44.0 \text{ g CO}_2}\right)\left(\dfrac{-393.5 \text{ kJ}}{\text{mol CO}_2}\right) = -315 \text{ kJ}$ (Data from Table 17.2)

343. We must start with a balanced equation for the reaction. We may then write in parentheses under each formula the enthalpy of formation, taken from Table 17.2, and multiply each enthalpy by the corresponding number of moles in the balanced equation. Remember that, by definition, ΔH_f° for any element in its standard state is 0.

$$2 \text{ Al} + \text{Fe}_2\text{O}_3 \quad \rightarrow \quad 2 \text{ Fe} + \text{Al}_2\text{O}_3$$
$$n\left(\Delta H_f^\circ\right) \quad 2(0) \quad 1(-822.2 \text{ kJ}) \quad 2(0) \quad 1(-1670 \text{ kJ})$$

Then ΔH° of the reaction is given by

$$\Delta H^\circ = (\text{sum of } \Delta H_f^\circ \text{ of products}) - (\text{sum of } \Delta H_f^\circ \text{ of reactants})$$
$$= -1670 \text{ kJ} - (-822.2 \text{ kJ}) = -848 \text{ kJ}$$

ΔH° for the reduction of Fe_2O_3: $\Delta H^\circ = -848$ kJ/mol

344. $C_6H_{12}O_6 + 6 O_2 \rightarrow 6 CO_2 + 6 H_2O$

$$\Delta H = 6 \Delta H_f^\circ (CO_2) + 6 \Delta H_f^\circ (H_2O) - \Delta H_f^\circ(C_6H_{12}O_6) = -2816 \text{ kJ}$$

$$\Delta H_f^\circ(C_6H_{12}O_6) = [6(-393.5) + 6(-285.9) - (-2816)] \text{ kJ}$$
$$= -1260 \text{ kJ} \qquad\qquad \text{(Data from Table 17.2)}$$

Since 1 mol of $C_6H_{12}O_6$ was involved, $\Delta H_f^\circ = -1260$ kJ/mol.

345. $\Delta H = 4\Delta H_f^\circ(CO_2) + 6\Delta H_f^\circ(H_2O) - 2\Delta H_f^\circ(C_2H_6) - 7\Delta H_f^\circ(O_2) = -3119$ kJ

$$\Delta H_f^\circ(C_2H_6) = \frac{[4(-393.5) + 6(-285.9) - (-3119)] \text{ kJ}}{2 \text{ mol}} = -85.2 \text{ kJ/mol}$$

346. $C_2H_4 + 3O_2 \rightarrow 2CO_2 + 2H_2O(l)$

$$\Delta H = 2\Delta H_f^\circ(CO_2) + 2\Delta H_f^\circ(H_2O) - \Delta H_f^\circ(C_2H_4)$$
$$= 2(-393.5 \text{ kJ}) + 2(-285.9 \text{ kJ}) - (51.9 \text{ kJ})$$
$$= -1411 \text{ kJ}$$

347. $\Delta H = 2\Delta H_f^\circ(HF) - 2\Delta H_f^\circ(HCl) = -353$ kJ

$$\Delta H_f^\circ(HCl) = \frac{[2(-269) - (-353)] \text{ kJ}}{2 \text{ mol}} = -92.5 \text{ kJ/mol}$$

348. **(A)** i (In iii, CO_2 is not formed from its elements.) **(B)** i (In ii, the C does not react as much as possible with O_2.) **(C)** iii **(D)** ii

349. Heat $= mc\Delta t = (500 \text{ g})(4.2 \text{ J/g}\cdot°C)[(26.25 - 25.08)°C] = 2.5 \times 10^3 \text{ J} = 2.5$ kJ

$$(400.0 \text{ mL})\left(\frac{0.200 \text{ mol}}{1000 \text{ mL}}\right) = 0.0800 \text{ mol acid} \quad (100.0 \text{ mL})\left(\frac{0.800 \text{ mol}}{1000 \text{ mL}}\right) = 0.0800 \text{ mol base}$$

$$\frac{2.5 \text{ kJ}}{0.0800 \text{ mol}} = 31 \text{ kJ/mol}$$

350. The complete combustion of a hydrocarbon involves the formation of CO_2 and H_2O.

$$C_2H_2(g) + \tfrac{5}{2}O_2(g) \rightarrow 2CO_2(g) \;+\; H_2O(l) \qquad \Delta H° = -1299 \text{ kJ}$$
$$n(\Delta H_f^\circ) \quad x \qquad\quad 0 \qquad\quad 2(-393.5 \text{ kJ}) \;\; -285.9 \text{ kJ}$$

Thus, $-1299 \text{ kJ} = [2(-393.5 \text{ kJ}) + (-285.9 \text{ kJ})] - x$

Solving, $x = \Delta H_f^\circ(C_2H_2) = +226$ kJ/mol.

351. The enthalpy of sublimation is merely the enthalpy of formation of $I_2(g)$.

$$\Delta H_{sub} = 62.43 \text{ kJ/mol}$$

352. $\dfrac{-92.4 \text{ kJ}}{2 \text{ mol}} = -46.2 \text{ kJ/mol}$

353. (1) $$C_2H_6 + \tfrac{7}{2}O_2 \rightarrow 2CO_2 + 3H_2O \quad \Delta H = -1541 \text{ kJ}$$

(2) $$C_2H_4 + 3O_2 \rightarrow 2CO_2 + 2H_2O \quad \Delta H = -1411 \text{ kJ}$$

(3) $$H_2 + \tfrac{1}{2}O_2 \rightarrow H_2O \quad \Delta H = -286 \text{ kJ}$$

(2) − (1) + (3) $$C_2H_4 + H_2 \rightarrow C_2H_6 \quad \Delta H = -156 \text{ kJ}$$

354.

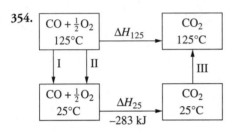

Figure 17.1

$$\Delta H_{125} = \Delta H_I + \Delta H_{II} + \Delta H_{25} + \Delta H_{III}$$

Using Figure 17.1, we have, per **mole** of CO:

$$\Delta H_I = (29.1 \text{ J/mol}\cdot\text{K})(-100 \text{ K}) = -2.91 \text{ kJ}$$
$$\Delta H_{II} = (0.500 \text{ mol})(29.5 \text{ J/mol}\cdot\text{K})(-100 \text{ K}) = -1.48 \text{ kJ}$$
$$\Delta H_{III} = (37.5 \text{ J/mol}\cdot\text{K})(+100 \text{ K}) = +3.75 \text{ kJ}$$
$$\Delta H_{25} = -283 \text{ kJ} \qquad\qquad \Delta H_{125} = -284 \text{ kJ}$$
$$(100 \text{ g CO})\left(\frac{1 \text{ mol}}{28.0 \text{ g}}\right)\left(\frac{-284 \text{ kJ}}{\text{mol CO}}\right) = -1010 \text{ kJ}$$

355. Let $x =$ number of moles of CO_2 produced; then $1.000 - x =$ number of moles of CO produced.

$$x\,(-393.5 \text{ kJ}) + (1.000 - x)(-110.5 \text{ kJ}) = -241 \text{ kJ}$$
$$x = 0.460 \text{ mol } CO_2 \qquad\qquad 1.000 - x = 0.540 \text{ mol CO}$$

356. $\text{Entropy} = \dfrac{\text{heat}}{\text{absolute temperature}}$ $\qquad$ $\text{Heat capacity} = \dfrac{\text{heat}}{\text{temperature change}}$

Since a temperature *change* is the same on the Kelvin and Celsius scales, either unit may be used in writing heat capacity.

357. (A) The reaction will just be spontaneous when $\Delta G = 0$.

$$\Delta G = \Delta H - T\Delta S = 0 \qquad \Delta H = T\Delta S \qquad T = \frac{\Delta H}{\Delta S} = \frac{100 \text{ kJ}}{0.050 \text{ kJ/K}} = 2000 \text{ K}$$

(B) $\Delta H = \Delta E + \Delta(PV) = \Delta E + \Delta(nRT) = \Delta E + (\Delta n)RT$

$= (-3000 \text{ J}) + (-1 \text{ mol})(8.314 \text{ J/mol} \cdot \text{K})(298 \text{ K}) = -5478 \text{ J}$

$\Delta G = \Delta H - T\Delta S = -5478 \text{ J} - 298 \text{ K}(-10.0 \text{ J/K}) = -2498 \text{ J}$

Since ΔG is negative, the reaction will be spontaneous as written.

358. ΔH is negative (the bond energy is released) and ΔS is also negative (there is less randomness among the molecules than among the atoms).

359. At 400 K, the reaction is at equilibrium.

$$\Delta G = \Delta H - T\Delta S = 0 \quad \text{at } 400 \text{ K}$$

$$\Delta S = \frac{\Delta H}{T} = \frac{-40,000 \text{ J}}{400 \text{ K}} = -100 \text{ J/K}$$

360. $\Delta G° = \left[\Delta G_f°(CO_2) + 2\Delta G_f°(H_2O)\right] - \left[\Delta G_f°(CH_4) + 2\Delta G_f°(O_2)\right]$

$= -394.6 \text{ kJ} + 2(-237.2 \text{ kJ}) - (-50.8 \text{ kJ}) - 0 \text{ kJ} = -818.2 \text{ kJ}$

361. At 0°C, $\Delta G = 0$

$$\Delta S = \frac{\Delta H}{T} = \frac{6004 \text{ J/mol}}{273 \text{ K}} = 22.0 \text{ J/mol} \cdot \text{K}$$

362. Per mol of N_2 used:

$$\Delta H° = 2\Delta H_f° = (2 \text{ mol})(33.8 \text{ kJ/mol}) = 67.6 \text{ kJ}$$

$$\Delta G° = 2\Delta G_f° = (2 \text{ mol})(51.9 \text{ kJ/mol}) = 103.8 \text{ kJ}$$

$$\Delta S° = (\Delta H° - \Delta G°)/T = (67.6 \text{ kJ} - 103.8 \text{ kJ}) / (298 \text{ K}) = -0.12 \text{ kJ/K}$$

Per 100 g of N_2:

$$\Delta S° = (100 \text{ g N}_2)\left(\frac{1 \text{ mol N}_2}{28.0 \text{ g N}_2}\right)\left(\frac{-0.12 \text{ kJ}}{\text{K} \cdot \text{mol N}_2}\right) = -0.43 \text{ kJ/K}$$

363. (A) $\Delta G° = \Delta G_{H_2O}° + \Delta G_{CO}° - \Delta G_{CO_2}°$

$\Delta G° = (-228.58 \text{ kJ} - 137.15 \text{ kJ}) - (-394.6 \text{ kJ}) = 28.9 \text{ kJ}$

(B) $\Delta G = \Delta G° + RT \ln Q$

$= (28.9 \text{ kJ}) + (8.314 \times 10^{-3} \text{ kJ/K})(298.1 \text{ K})\left[\ln \frac{P(H_2O)P(CO)}{P(H_2)P(CO_2)}\right]$

$= \left[28.9 + 2.478 \ln \frac{(0.02000)(0.01000)}{(10.00)(20.00)}\right] \text{kJ} = (28.9 - 34.23) \text{ kJ} = -5.3 \text{ kJ}$

Note that the reaction, although not possible under standard conditions, becomes possible ($\Delta G < 0$) under this set of experimental conditions. In this case, Q was evaluated in terms of partial pressures, since $\Delta G°$ was defined in terms of a standard state of 1 atm for each substance. In general, Q must be expressed in terms of the same concentration measure used to define the standard states.

364. (A) $\Delta G = \Delta H - T\Delta S = (-15,000 \text{ J/mol}) - (298 \text{ K})(-7.2 \text{ J/mol} \cdot \text{K}) = -12.9 \text{ kJ/mol}$

Since the value of ΔG is negative, the reaction may proceed spontaneously.

(B) $\Delta G° = \Delta H° - T\Delta S° = \Delta E° + \Delta(PV) - T\Delta S°$

Assuming ideal gas behavior,
$$\Delta G° = \Delta E° + (\Delta n)RT - T\Delta S°$$

Using the value of $R = 8.314 \text{ J/mol} \cdot \text{K}$, and the fact that 2 mol of gas (D) is produced from 3 mol (2 A + B),

$$(\Delta n)RT = (-1 \text{ mol})(8.314 \text{ J/mol} \cdot \text{K})(298 \text{ K}) = -2480 \text{ J}$$

$$\Delta G° = -4500 \text{ J} + (-2480 \text{ J}) - (298 \text{ K})(-10.5 \text{ J/K}) = -3.85 \text{ kJ}$$

Since the value of $\Delta G°$ is negative, the indicated reaction is spontaneous.

365. $CaCO_3 \rightarrow CaO + CO_2 \quad \Delta G = \Delta H - T\Delta S$

Increasing T makes the second term relatively more important. Since ΔS is positive for the decomposition reaction, the increasing magnitude of $T\Delta S$ causes ΔG to become negative at high temperature, signaling the spontaneous decomposition of $CaCO_3$.

Chapter 18: Chemical Kinetics

366. Rate = $k[A]^m[B]^n$ The rate is a variable, proportional to the concentrations and to the rate constant. The rate constant, k, is a number that depends on the temperature only for a given reaction and not on the concentrations.

Table 18.1   Concentration as a Function of Time*

Order (n) with Respect to A	Rate Equation in Terms of Reactant Concentration	Rate Equation in Terms of Product Concentration	Units of k
0	$[A_0] - [A] = kt$	$[X] = kt$	$\text{mol} \cdot \text{L}^{-1} \cdot \text{s}^{-1}$
1	$\begin{cases} \ln[A_0] - \ln[A] = kt \\ \text{or} \quad \ln\dfrac{[A]}{[A_0]} = -kt \end{cases}$	$\begin{cases} \ln[A_0] - \ln([A_0]-[X]) = kt \\ \text{or} \quad \dfrac{[A_0]-[X]}{[A_0]} = -kt \end{cases}$	s^{-1}
2	$\dfrac{1}{[A]} - \dfrac{1}{[A_0]} = kt$	$\dfrac{1}{[A_0]-[X]} - \dfrac{1}{[A_0]} = kt$	$\text{L} \cdot \text{mol}^{-1} \cdot \text{s}^{-1}$
3	$\dfrac{1}{[A]^2} - \dfrac{1}{[A_0]^2} = 2kt$	$\dfrac{1}{([A_0]-[X])^2} - \dfrac{1}{[A_0]^2} = 2kt$	$\text{L}^2 \cdot \text{mol}^{-2} \cdot \text{s}^{-1}$

* Rate = $k[A]^n$.

[A] = concentration of a reactant at a given time, t (s).
[A_0] = initial concentration of the reactant.
[X] = concentration of a product at a given time, t (s).
Each ln may be replaced by 2.30 log.

367. Those with integral orders are: Rate $= k[A][B]$ Rate $= k[A]^2$ Rate $= k[B]^2$

368. (A) From the coefficients in the balanced equation, $\Delta n(N_2) = -\frac{1}{2}\Delta n(NH_3)$. Therefore,

$$-\frac{\Delta[N_2]}{\Delta t} = \frac{1}{2}\frac{\Delta[NH_3]}{\Delta t} = 1.0 \times 10^{-4}\ \text{mol} \cdot \text{L}^{-1} \cdot \text{s}^{-1}$$

(B) Similarly, $-\dfrac{\Delta[H_2]}{\Delta t} = \dfrac{3}{2}\dfrac{\Delta[NH_3]}{\Delta t} = 3.0 \times 10^{-4}\ \text{mol} \cdot \text{L}^{-1} \cdot \text{s}^{-1}$

Note the minus signs. As NH_3 is produced, the elements are used up.

369. In each case, we write out the full rate equation and find the units of k that will satisfy the equation.

(A) $-\dfrac{\Delta[A]}{\Delta t} = k$ Units of k = units of $\dfrac{[A]}{t} = \dfrac{\text{mol/L}}{\text{s}} = \text{mol} \cdot \text{L}^{-1} \cdot \text{s}^{-1}$

Note that the units of $\Delta[A]$, the change in concentration, are the same as the units of $[A]$ itself; similarly for Δt.

(B) $-\dfrac{\Delta[A]}{\Delta t} = k[A]$ or $k = -\dfrac{1}{[A]}\dfrac{\Delta[A]}{\Delta t}$

Units of $k = \left(\dfrac{1}{\text{mol/L}}\right)\left(\dfrac{\text{mol/L}}{\text{s}}\right) = \text{s}^{-1}$

First order reactions are the only reactions for which k has the same numerical value regardless of the units used for expressing the concentrations of the reactants or products.

(C) $-\dfrac{\Delta[A]}{\Delta t} = k[A]^2$ $\Bigg|$ $-\dfrac{\Delta[A]}{\Delta t} = k[A][B]$

$k = -\dfrac{1}{[A]^2}\dfrac{\Delta[A]}{\Delta t}$ $\Bigg|$ $k = -\dfrac{1}{[A][B]}\dfrac{\Delta[A]}{\Delta t}$

Units of $k = \left(\dfrac{1}{(\text{mol/L})^2}\right)\dfrac{\text{mol/L}}{\text{s}} = \text{L} \cdot \text{mol}^{-1} \cdot \text{s}^{-1}$

Note that the units of k depend on the *total* order of the reaction, not on the way the total order is composed of the orders with respect to different reactants.

(D) $-\dfrac{\Delta[A]}{\Delta t} = k[A]^3$ or $k = -\dfrac{1}{[A]^3}\dfrac{\Delta[A]}{\Delta t}$

Units of $k = \left(\dfrac{1}{(\text{mol/L})^3}\right)\left(\dfrac{\text{mol/L}}{\text{s}}\right) = \text{L}^2 \cdot \text{mol}^{-2} \cdot \text{s}^{-1}$

370. Most product, **(B)**. Since there is more reactant present, there will be more product produced per unit time. The highest rate, the greatest change in *concentration* per unit time, is **(C)**, simply because the vessel with the greatest concentration will also experience the greatest change in concentration per unit time.

371. (A) It is apparent from the data on the left that, all other factors the same, the initial rate of the reaction is proportional to the initial concentration of A. The reaction is first order with respect to A. Note that no conclusions about the order with respect to B can be drawn from the data on the left-hand side of the table. The data on the right show that the initial rate is proportional to the square of the initial concentration of B, and therefore the reaction is second order with respect to B. The reaction is third order overall. The numerical value of the rate constant, k, may be obtained by substituting data for each experiment into the rate law expression:

$$\text{Rate} = k[A]^m[B]^n = k[A][B]^2$$

Thus, for the data in the first line,

$$1.2 \times 10^{-2}\,\text{M/s} = k(1.0\,\text{M})(1.0\,\text{M})^2 \quad \text{so} \quad k = 1.2 \times 10^{-2}\,\text{M}^{-2} \cdot \text{s}^{-1}$$

(B) From the first statement it may be deduced that the reaction is first order with respect to A. Since doubling both concentrations causes only a doubling of the rate, which is caused by the doubling of the A concentration, the doubling of the B concentration has had no effect on the rate; the reaction is zero order with respect to B.

372. Initial rate $= (0.137\,\text{M}^{-1} \cdot \text{s}^{-1})(0.050\,\text{M})^2 = 3.4 \times 10^{-4}\,\text{M} \cdot \text{s}^{-1}$

$$\frac{0.030(0.050\,\text{M})}{3.4 \times 10^{-4}\,\text{M} \cdot \text{s}^{-1}} = 4.4\,\text{s}$$

(This answer assumes that the initial rate does not drop appreciably as small amounts of the reactants are used up.)

373. This reaction is pseudo first order. No matter how complete the reaction of the complex ion, the concentration of the water changes very little, and therefore it is impossible to determine the order with respect to water.

374. Since the concentrations of the two reactants start equal and remain equal throughout the reaction, the reaction can be treated as a simple second order reaction.

$$\frac{1}{[A]} = kt + \frac{1}{[A_0]} = \frac{1.0 \times 10^{-2}\,\text{L}}{\text{mol} \cdot \text{s}}(100\,\text{s}) + \frac{1\,\text{L}}{0.100\,\text{mol}} = 11\,\text{L/mol}$$

$$[A] = 0.091\,\text{M}$$

375. Since the initial concentration of B is 6.00 M, and throughout the reaction the B concentration cannot possibly fall below 5.90 M, it is useful to treat the B concentration as a constant. The rate expression can be written in the form

$$\text{Rate} = 6.00k[A]$$

Treating $6.00k$ as a new rate constant, k', one can use the first order rate law equation in the form

$$\ln[A] = \ln[A_0] - 6.00kt$$
$$\ln[A] = -2.30 - (3.0 \times 10^{-2})(100) = -2.30 - 3.0 = -5.3$$
$$[A] = 5.0 \times 10^{-3}\,\text{M}$$

376. The data may be tested to see if they fit any of the equations of Table 18.1:

$$\text{1st order} \qquad \ln\frac{[A]}{[A_0]}=-kt \qquad \ln\frac{10^{-3}}{10^{-2}}\neq-9.0(100)$$

$$\text{2nd order} \qquad \frac{1}{[A]}-\frac{1}{[A_0]}=kt \qquad \frac{1}{10^{-3}}-\frac{1}{10^{-2}}=9.0(100)$$

The reaction may be seen to be second order, since the data fit that equation. Since the second order equation fits the data, it is not necessary to calculate the values for the third or zero order equations.

377. Inspection of the data is sufficient to see that $[A] - [A_0]$ is proportional to t, and therefore the reaction is zero order, with $k = 0.010$ M/min.

378. **(A)** rate $= k[A] = (5.00 \times 10^{-5}\,\text{s}^{-1})(1.00\,\text{M}) = 5.00 \times 10^{-5}$ M/s

(B) $\ln\dfrac{[A]}{[A_0]} = -kt = -(5.00 \times 10^{-5}\,\text{s}^{-1})(3600\,\text{s}) = -0.180$

$[A] = 0.835$ M

Rate $= k[A] = (5.0 \times 10^{-5}\,\text{s}^{-1})(0.835\,\text{M}) = 4.2 \times 10^{-5}$ M/s

379. **(A)** $t_{1/2} = \dfrac{\ln 2}{k} = \dfrac{0.693}{7.1 \times 10^{-5}\,\text{s}^{-1}} = 9.8 \times 10^{3}\,\text{s} = 2.7$ h

(B) $k = \dfrac{\ln 2}{t_{1/2}} = \dfrac{0.693}{2.50\,\text{h}}\left(\dfrac{1\,\text{h}}{3600\,\text{s}}\right) = 7.70 \times 10^{-5}\,\text{s}^{-1}$

380. At any time, t, the fraction of sucrose remaining is $[A]/[A_0]$.

$$\ln\frac{[A]}{[A_0]} = -kt = -\left(\frac{\ln 2}{t_{1/2}}\right)t = -\left(\frac{0.693}{3.33\,\text{h}}\right)(9.00\,\text{h}) = -1.87$$

$$\frac{[A]}{[A_0]} = 0.154 = \text{fraction remaining}$$

381. For each mole of azomethane that decomposes, 2 mol of gaseous products are obtained. Since in a constant volume at constant temperature the pressure of a gas is directly proportional to the number of moles of gas, the total pressure in the flask will increase as the reaction proceeds. The pressure increase is equal to the partial pressure of N_2 (or C_2H_6). Since the reaction is first order, the rate equation in terms of pressure (concentration) of N_2 is as follows (Table 18.1):

$$\ln\frac{P_0 - P_{N_2}}{P_0} = -kt$$

A plot of the logarithm term against time, t, gives a straight line with a slope equal to $-k$. In this case, $k = 1.30 \times 10^{-2}$/min. The half-life is given by $0.693/k = t_{1/2}$; therefore, $t_{1/2} = 53.3$ min.

382. (A) A catalyst lowers the activation energy; therefore, more molecules in the sample have sufficient energy to react. **(B)** A temperature increase causes a greater fraction of the molecules to have an energy at least equal to the activation energy. **(C)** A concentration increase causes a greater number of collisions per second, therefore a greater number of effective collisions per second.

383. (A) y **(B)** $x + y$

384. An activated complex by definition is unstable and has no independent existence. ΔH_f° is probably positive. The bond orders are apt to be unusual, and the bond lengths and angles are likely to differ from those of analogous stable molecules.

385. See Figure 18.2.

$$\Delta H = 163 \text{ kJ} - 184 \text{ kJ} = -21 \text{ kJ}$$

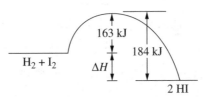

Figure 18.2

386. An unstable intermediate is an actual chemical species (which perhaps can be stabilized under different reaction conditions). It has normal bond orders for its atoms. It represents a minimum on the potential energy curve (Figure 18.3), albeit a small minimum. The activated complex is the postulated species that has maximum energy during the conversion from reactants to products. No matter which way the bond lengths and strengths vary, stabilization results. The bond orders of the atoms of an activated complex are sometimes unusual.

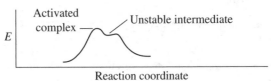

Figure 18.3

387. Since this is a single-step reaction, the order with respect to each reactant is equal to the molecularity of the reactant.

$$\text{Rate} = k[\text{A}][\text{B}]$$

388. No. An elementary process must have a rate law with orders equal to its molecularities, which are necessarily integers.

389. The first elementary process is first order with respect to A and with respect to B and, therefore, second order overall:

$$\text{Rate} = k[A][B]$$

Although the second elementary process involves only one reactant, X, it is a bimolecular process, and therefore is second order:

$$\text{Rate} = k[X]^2$$

390. An accepted mechanism is one that fits all the data available. No mechanism can be proved, since it is always possible that further data will necessitate a change.

391. $\text{Rate} = k_2[N_2O_2][H_2]$

$$K_1 = \frac{[N_2O_2]}{[NO]^2} \quad \text{hence } [N_2O_2] = K_1[NO]^2$$

where K_1 is the equilibrium constant for the first step.

$$\text{Rate} = k_2 K_1[NO]^2[H_2] = k[NO]^2[H_2]$$

392. Combining the first reaction with twice the second (so the NO and NO_2 are eliminated) yields the following reaction:

$$2\ SO_2 + O_2 \rightarrow 2\ SO_3$$

That reaction proceeds only slowly in the absence of the NO and/or NO_2. The NO or NO_2 is a catalyst. It speeds up the conversion of SO_2 to SO_3 and is not changed permanently in the process.

Chapter 19: Equilibrium

393. $H_2O(s) + \text{heat} \rightleftarrows H_2O(l)$

The equilibrium shifts toward ice as you remove heat in an attempt to lower the temperature. (However, the temperature does not change until all the water freezes and the equilibrium is destroyed.)

394. Nothing. Le Châtelier's principle applies only to a system already at equilibrium.

395. (A) The equilibrium must shift to the right to use up some of the added nitrogen, thus lowering its total concentration and thereby reducing the stress. A new equilibrium is established. Note that the equilibrium does not shift back to the left because of the additional ammonia formed, since that was produced by the reaction itself. Note also that Le Châtelier's principle alone does not tell *how much* the equilibrium will be shifted. No matter how large the quantity of nitrogen added, some hydrogen will be present at the new equilibrium state.

(B) In the application of Le Châtelier's principle, heat energy may be treated as one of the reactants or products of the reaction. Heat must be added to raise the temperature of the system. Therefore in order to use up some of the added heat, the equilibrium must shift to the left.

(C) The reaction of a total of 4 mol of hydrogen and nitrogen would produce 2 mol of ammonia. In a given volume, fewer moles of gas exert less pressure; thus, as required by Le Châtelier's principle, if additional pressure is applied by lowering the volume, the equilibrium will shift to the right, in the direction of fewer moles of gas. (For an example, see Question 266.)

396. (A) Since the equation states that 2 mol of gaseous reactant(s) would produce 2 mol of gaseous product(s), changing the total pressure will not shift the equilibrium at all.

(B) The equilibrium would shift to the right. The increase in volume would decrease the pressure of each of the gases, but not of the carbon, which is a solid. The number of moles of gas would be increased by the shift to the right.

397. (A) $K = \dfrac{[NO_2][NO_2]}{[N_2O_4]}$ (B) $K = \dfrac{[NO_2]^2}{[N_2O_4]}$ (C) $K = \dfrac{[N_2O_4]}{[NO_2]^2}$ (D) $K = \dfrac{[N_2O_4]^{1/2}}{[NO_2]}$

The equilibrium constant expressions of parts (A) and (B) are the same because the chemical equations are the same. It is easy to see why the coefficients of the chemical equation become exponents in the equilibrium constant expression by writing out the equation and expression as in part (A).

The K in part (C) is the reciprocal of that in parts (A) and (B). The K in part (D) is the square root of that in part (C).

398. (A) $K = \dfrac{[Z]}{[X][Y]} = \dfrac{0.130}{(0.120)(0.250)} = 4.33$

(B) $K = \dfrac{0.0650}{(0.0600)(0.125)} = 8.67$

The value is not the same. The value of K is related to the *concentrations* of the reagents, not to their numbers of moles.

399. (A) The calculations in the following table were done in the order indicated by the superscript numbers:

	A	+	B	⇌	C	+	D
Initial:	^{1}0.800		^{1}1.60		^{1}0.000		^{1}0.000
Change:	^{3}−0.300		^{3}−0.300		^{2}+0.300		^{3}+0.300
Equilibrium:	^{4}0.500		^{4}1.30		^{1}0.300		^{4}0.300

$$K = \dfrac{[C][D]}{[A][B]} = \dfrac{(0.300)^2}{(0.500)(1.30)} = 0.138$$

(B) The terms in the second line are in that ratio, since it describes the changes made by the reaction. The terms in the other lines are not in that ratio.

400.

$$2\,A \quad + \quad B \quad \rightleftarrows \quad C \quad + \quad 2\,D$$

Initial:	1.10	1.40	0	0
Change:	−0.60	−0.30	+0.30	+0.60
Equilibrium	0.50	1.10	0.30	0.60

$$K = \frac{[C][D]^2}{[A]^2[B]} = \frac{(0.30)(0.60)^2}{(0.50)^2(1.10)} = 0.39$$

401. (A) The initial concentration of each reagent is 0.100 mol/2.00 L = 0.0500 mol/L. Let the equilibrium concentration of HI be $2x$ for convenience.

$$H_2 \quad + \quad I_2 \quad \rightleftarrows \quad 2\,HI$$

Initial:	0.0500	0.0500	0
Change:	−x	−x	+2x
Equilibrium:	0.0500 − x	0.0500 − x	2x

$$K = \frac{[HI]^2}{[H_2][I_2]} = \frac{(2x)^2}{(0.0500 - x)^2} = 50$$

Taking the square root of each side of the last equation yields

$$\frac{2x}{0.0500 - x} = 7.1$$
$$2x = 7.1(0.0500) - 7.1x$$
$$9.1x = 7.1(0.0500)$$
$$x = 0.039$$
$$[I_2] = 0.0500 - 0.039 = 0.011 \text{ mol/L}$$

In 2.00 L, there is, then, 0.022 mol I_2.

(B) $K = \dfrac{[CO_2][H_2]}{[CO][H_2O]} = \dfrac{x^2}{(0.0500 - x)^2} = 0.64$

Taking the square root of this expression yields

$$\frac{x}{0.0500 - x} = 0.80$$
$$x = 0.040 - 0.80x$$
$$1.80x = 0.040$$
$$x = 0.022 \text{ mol/L}$$

402. $PCl_5 \rightleftarrows PCl_3 + Cl_2$

$$K = 0.0410 = \frac{[PCl_3][Cl_2]}{[PCl_5]} = \frac{x^2}{0.100 - x}$$

$$x^2 = 0.00410 - 0.0410x$$

$$x^2 + 0.0410x - 0.00410 = 0$$

$$x = \frac{-0.0410 + \sqrt{(0.0410)^2 + 4(0.00410)}}{2} = 0.0467$$

moles $Cl_2 = (0.0467 \text{ mol/L})(5.00 \text{ L}) = 0.234 \text{ mol}$

403.

	$H_2(g)$	+	$S(s)$	$\rightleftarrows$	$H_2S(g)$
Initial:	0.20				0.00
Change:	$-x$				$+x$
Equilibrium:	$0.20 - x \approx 0.20$				x

$$K = \frac{[H_2S]}{[H_2]} = \frac{x}{0.20} = 6.8 \times 10^{-2} \qquad x = 1.4 \times 10^{-2} \text{ mol/L}$$

$$P(H_2S) = \left(\frac{n}{V}\right)RT = (1.4 \times 10^{-2} \text{ mol/L})(0.0821 \text{ L} \cdot \text{atm/mol} \cdot \text{K})(363 \text{ K}) = 0.42 \text{ atm}$$

404.

	N_2O_4	$\rightleftarrows$	$2NO_2$
Initial:	0.100		0.100
Change:	-0.025		$+0.050$
Equilibrium:	0.075		0.150

$$K = \frac{[NO_2]^2}{[N_2O_4]} = \frac{(0.150)^2}{0.075} = 0.30$$

405.

	$2XO$	$\rightleftarrows$	X_2O_4
Initial:	0.100		0.100
Change:	$+0.050$		-0.025
Equilibrium:	0.150		0.075

$$K = \frac{[X_2O_4]}{[XO_2]^2} = \frac{(0.075)}{(0.150)^2} = 3.3$$

406. $I_2 + I^- \rightleftarrows I_3^-$ $\qquad K = \dfrac{[I_3^-]}{[I_2][I^-]}$

At equilibrium,

$[I_2] = 1.30 \times 10^{-3}\,M$ $\qquad\qquad [I_3^-] = (0.0492\ M) - (0.00130\ M) = 0.0479\ M$

$[I^-] = (0.100\ M) - (0.0479\ M) = 0.0521\ M$

$K = \dfrac{0.0479}{(1.30 \times 10^{-3})(0.0521)} = 707$

407. For the reaction

$$a\,A(g) + b\,B(s) \rightleftarrows c\,C(g) + d\,D(s) \qquad\qquad K = \dfrac{[C]^c}{[A]^a} \qquad K_p = \dfrac{P_C^c}{P_A^a}$$

(The activities of the solids are each equal to 1.) The concentrations of the gases, in mol/L, are given by

$$\dfrac{n}{V} = \dfrac{P}{RT} \qquad\qquad \text{Thus:} \quad K = \dfrac{(P_C/RT)^c}{(P_A/RT)^a} = K_p(RT)^{a-c} = K_p(RT)^{-\Delta n}$$

$$K_p = K(RT)^{\Delta n}$$

408. Two different values of $\Delta G°$ will be obtained, but the first refers to the standard state in which all reactants and products are 1 M and the other refers to the standard state in which all reactants and products are at 1 atm.

409. $\Delta G = 0$ for any system at equilibrium.

$$\Delta G° = -RT \ln K = -(8.31\ \text{J/mol} \cdot \text{K})(300\ \text{K})(2.30) = -5.73\ \text{kJ/mol}$$

410. (A) $\Delta G° = \Delta H° - T\,\Delta S° = (-29.8\ \text{kJ}) - (298\ \text{K})(-0.100\ \text{kJ/K}) = 0.00$

$\qquad\qquad \Delta G° = -RT \ln K = 0.00$

$\qquad\qquad \ln K = 0.00$

$\qquad\qquad K = 1.0$

(B) $K = \dfrac{[HI]^2}{[H_2][I_2]} = \dfrac{(1.02 \times 10^{-2})^2}{(8.62 \times 10^{-4})(2.63 \times 10^{-3})} = 45.9$

$$\Delta G° = -RT \ln K = -(8.31\ \text{J/K})(763\ \text{K})(\ln 45.9) = -24.3 \times 10^3\ \text{J} = -24.3\ \text{kJ}$$

Chapter 20: Acids and Bases

411. **(A)** The sodium acetate shifts the ionization reaction to the left. Note that the sodium acetate does not react with the acetic acid to produce other products.

$$HC_2H_3O_2 + H_2O \rightleftharpoons C_2H_3O_2^- + H_3O^+$$

(B)

acid		neutral	basic		
HCl HC$_2$H$_3$O$_2$	NH$_4$Cl	NaCl	(NH$_3$ + NH$_4$Cl)	NH$_3$	NaOH

Increasing pH →

NH$_4$Cl is acidic because of the hydrolysis of NH$_4^+$. The buffer solution is less basic than ammonia alone because the NH$_4^+$ represses the ionization of NH$_3$ in the buffer.

412. **(A)** $HC_2H_3O_2 + H_2O \rightarrow H_3O^+ + C_2H_3O_2^-$ < 2%

(B) $C_2H_3O_2^- + H_2O \rightarrow HC_2H_3O_2 + OH^-$ < 2%

(C) $C_2H_3O_2^- + H_3O^+ \rightarrow HC_2H_3O_2 + H_2O$ > 98%

413. In each case the conjugate base is derived from the acid by the loss of a proton. **(A)** CN$^-$ **(B)** CO$_3^{2-}$ **(C)** N$_2$H$_4$

414. **(A)**

$$HX \ + \ H_2O \ \rightleftharpoons \ H_3O^+ \ + \ X^-$$

acid   base   acid   base

conjugates

conjugates

(B)

$$NH_3 + H_2O \rightleftharpoons NH_4^+ + OH^-$$

$$K_b = \frac{[NH_4^+][OH^-]}{[NH_3]} = \frac{(0.00134)^2}{(0.100 - 0.00134)} = 1.82 \times 10^{-5}$$

415. **(A)** $NH_2CH_2COO^-$ **(B)** $^+NH_3CH_2COOH$

416.

$$NH_3 + H_2O \rightleftharpoons NH_4^+ + OH^-$$

$$K_b = \frac{[NH_4^+][OH^-]}{[NH_3]}$$

Ionization of the base produces equal concentrations of hydroxide and ammonium ions:

$$[NH_4^+] = [OH^-] = 1.34 \times 10^{-3} \text{ M}$$

$[NH_3] = 0.100 - (1.34 \times 10^{-3}) = 0.099$ M

$$K_b = \frac{(1.34 \times 10^{-3})^2}{(0.099)} = 1.8 \times 10^{-5}$$

417. $HC_2H_3O_2 + H_2O \rightleftharpoons H_3O^+ + C_2H_3O_2^-$ $K_a = \dfrac{[H_3O^+][C_2H_3O_2^-]}{[HC_2H_3O_2]} = 1.80 \times 10^{-5}$

At equilibrium, let $[H_3O^+] = x$, then $[C_2H_3O_2^-] = x$ and $[HC_2H_3O_2] = 0.200 - x$

$$\frac{x^2}{0.200 - x} = 1.80 \times 10^{-5}$$

Neglecting x when subtracted from 0.200 yields

$$x = 1.90 \times 10^{-3} \, M = [H_3O^+]$$

418. $NH_3 + H_2O \rightleftharpoons NH_4^+ + OH^-$

$$K_b = \frac{[NH_4^+][OH^-]}{[NH_3]} = \frac{x^2}{0.10 - x} \approx \frac{x^2}{0.10} = 1.8 \times 10^{-5}$$

$$x = 1.3 \times 10^{-3} = [OH^-]$$

419. Let x = total concentration of acetic acid.

$$[H^+] = [C_2H_3O_2^-] = 3.5 \times 10^{-3} \qquad\qquad [HC_2H_3O_2] = x - 3.5 \times 10^{-3}$$

$$\frac{[H^+][C_2H_3O_2^-]}{[HC_2H_3O_2]} = \frac{(3.5 \times 10^{-3})^2}{x - 3.5 \times 10^{-3}} = 1.8 \times 10^{-5}$$

$$x = 6.8 \times 10^{-1} \, M$$

420. The number 10.92 has four significant figures; 0.92 has two. Since the integral portion of a logarithm determines the power of 10 only, each hydronium ion concentration should be reported to 2 significant digits—the same as the number of decimal digits in the pH.

antilog $(-10.92) = 10^{-10.92} = 1.2 \times 10^{-11}$ antilog $(-0.92) = 10^{-0.92} = 1.2 \times 10^{-1}$

421. (A) $K_w = [H_3O^+][OH^-]$

$$[OH^-] = 10^{-2} \qquad\qquad [H_3O^+] = 10^{-14}/10^{-2} = 10^{-12} \quad \text{thus,} \quad pH = 12$$

(B) $pH = -\log(6.0 \times 10^{-8}) = 7.22$

422. (A) $pH = -\log[H_3O^+] = 11.73$

$$[H_3O^+] = 1.9 \times 10^{-12}$$

$$K_w = [H_3O^+][OH^-] = 1.0 \times 10^{-14}$$

$$[OH^-] = \frac{1.0 \times 10^{-14}}{1.9 \times 10^{-12}} = 5.3 \times 10^{-3} \, M$$

(B) The hydroxide ion concentration from the autoionization of water is negligible compared with that provided from the NaOH, and the solution consists of 0.100 M Na^+ and 0.100 M OH^-. Hence, $[OH^-] = 0.100$ M. In any dilute aqueous solution,

$$K_w = [H_3O^+][OH^-] = 1.0 \times 10^{-14}$$

In this solution, $1.0 \times 10^{-14} = [H_3O^+](0.100)$; thus, $[H_3O^+] = 1.0 \times 10^{-13}$.

423. (A) 1.0   **(B)** 1.4×10^{-14}

424. (A) $[OH^-] = 2.0 \times 10^{-3}$ $[H_3O^+] = 5.0 \times 10^{-12}$ thus,   pH = 11.30

(B) $[H_3O^+] = [OH^-] = 1.0 \times 10^{-7}$ so   pH = 7   (NaCl is neither acidic nor basic.)

425. (A) pH = 4.500   so $[H_3O^+] = 3.16 \times 10^{-5}$ $HA + H_2O \rightarrow H_3O^+ + A^-$

	H_2O +	HA	$\rightarrow H_3O^+$	+	A^-
Initial:		0.10	0.00		0.00
Change:		-3.16×10^{-5}	3.16×10^{-5}		3.16×10^{-5}
Equilibrium:		0.10	3.16×10^{-5}		3.16×10^{-5}

$$K_a = \frac{[H_3O^+][A^-]}{[HA]} = \frac{(3.16 \times 10^{-5})^2}{0.10} = 1.0 \times 10^{-8}$$

(B) pH = 10.500   and   pOH = 3.500   so $[OH^-] = 3.16 \times 10^{-4}$

$$B + H_2O \rightleftharpoons BH^+ + OH^- \qquad K_b = \frac{[OH^-][BH^+]}{[B]} = \frac{(3.16 \times 10^{-4})^2}{(0.10)} = 1.0 \times 10^{-6}$$

426. The diluted HCl is 0.10 M; the pH is 1.00. (With added strong acid present, the H_3O^+ from ionization of the weak acid, $HC_2H_3O_2$, is insignificant.)

427. (A) $HCl + H_2O \rightarrow H_3O^+ + Cl^-$ (100% ionization)

0.010 M H_3O^+ pH = 2.00

(B) The salt contains acetate ions, which react with hydronium ions:

$$H_3O^+ + C_2H_3O_2^- \rightarrow HC_2H_3O_2 + H_2O$$

Assume complete reaction of the hydronium ion, yielding 0.010 M $HC_2H_3O_2$ and leaving 0.010 M $C_2H_3O_2^-$ excess:

$$HC_2H_3O_2 + H_2O \rightleftharpoons C_2H_3O_2^- + H_3O^+$$

$$K_a = \frac{[C_2H_3O_2^-][H_3O^+]}{[HC_2H_3O_2]} = \frac{(0.010)[H_3O^+]}{(0.010)} = 1.80 \times 10^{-5}$$

$[H_3O^+] = 1.8 \times 10^{-5}$ so pH $= 4.74$ and ΔpH $= 4.74 - 2.00 = 2.74$

428. Addition of OH$^-$ does not shift an equilibrium toward un-ionized base, as it would with a weak base and its conjugate.

429. **(A)**, **(B)**, **(G)**, and **(H)**. The solution in **(G)** contains equimolar quantities NH$_4^+$ and NH$_3$. The NH$_4^+$ resulted from the reaction

$$H^+Cl^- + NH_3 \rightarrow NH_4^+ + Cl^-$$

The NH$_3$ was present in excess. A similar situation occurs in **(H)**.

430.

	HA	+	H$_2$O	→	H$_3$O$^+$	+	A$^-$
Initial:	0.10				0.00		0.20
Change:	$-x$				x		x
Equilibrium:	0.10 $- x$				x		0.20 $+ x$
	≈ 0.10						≈ 0.20

$\dfrac{x(0.20)}{(0.10)} = 1.0 \times 10^{-7}$ thus, $x = 5.0 \times 10^{-8} = [H_3O^+]$ and pH $= 7.30$

431. **(A)** 0.10 mol NaC$_2$H$_3$O$_2$. **(B)** 0.10 mol NaOH. The 0.10 mol of NaOH will react with 0.10 mol of HC$_2$H$_3$O$_2$, producing a solution equivalent to that produced in part **(A)**. **(C)** 0.10 mol HCl. The HCl will react C$_2$H$_3$O$_2^-$ (producing 0.10 mol of HC$_2$H$_3$O$_2$) and yielding the same solution as produced in **(A)** except with NaCl added. **(D)** 0.10 mol HC$_2$H$_3$O$_2$ **(E)** 0.20 mol HC$_2$H$_3$O$_2$. Half of this acid will neutralize the NaOH, yielding the same solution as produced in part **(A)**. Note that this combination of reactants is the same as in part **(B)**.

432. $[H_3O^+] = 1.0 \times 10^{-7}$ $\qquad K_a = \dfrac{[H_3O^+][C_2H_3O_2^-]}{[HC_2H_3O_2]} = 1.80 \times 10^{-5}$

$\dfrac{[HC_2H_3O_2]}{[C_2H_3O_2^-]} = \dfrac{1.0 \times 10^{-7}}{1.80 \times 10^{-5}} = 5.6 \times 10^{-3}$

It is possible for an acid to exist at pH 7 if, in the same solution, some base is present—in this case, C$_2$H$_3$O$_2^-$ (in 180 times the concentration).

433. $NH_3 + H_2O \rightleftharpoons NH_4^+ + OH^-$

$K_b = \dfrac{[NH_4^+][OH^-]}{[NH_3]} = \dfrac{(0.100)x}{(0.050)} = 1.8 \times 10^{-5}$ thus, $x = [OH^-] = 9.0 \times 10^{-6}$

434. $HC_2H_3O_2 + H_2O \rightleftharpoons C_2H_3O_2^- + H_3O^+$

$$K_a = \frac{[H_3O^+][C_2H_3O_2^-]}{[HC_2H_3O_2]} = \frac{x(0.100)}{(0.080)} = 1.80 \times 10^{-5} \quad \text{thus,} \quad x = 1.4 \times 10^{-5}$$

435. The acid and base react, leaving 0.0500 M $HC_2H_3O_2$ and 0.0500 M $C_2H_3O_2^-$ (in 100 mL).

$$K_a = \frac{[H_3O^+][C_2H_3O_2^-]}{[HC_2H_3O_2]} = 1.80 \times 10^{-5} \qquad [H_3O^+] = 1.80 \times 10^{-5} \quad \text{thus,} \quad pH = 4.74$$

436. $HC_2H_3O_2 + H_2O \rightarrow H_3O^+ + C_2H_3O_2^- \quad K_a = \dfrac{[H_3O^+][C_2H_3O_2^-]}{[HC_2H_3O_2]} = 1.80 \times 10^{-5}$

(A) $[H_3O^+] = [C_2H_3O_2^-] = x \qquad [HC_2H_3O_2] = 0.10 - x \approx 0.10$

$$\frac{x^2}{0.10} = 1.8 \times 10^{-5} \quad \text{thus,} \quad x = 1.34 \times 10^{-3} = [H_3O^+] \quad \text{and} \quad pH = 2.87$$

(B) Half the acid is neutralized and half remains. Both solutes are now in 75.0 mL of solution.

$$[H_3O^+] = y \qquad [C_2H_3O_2^-] = \frac{(0.050\ M)(50.0\ mL)}{(75.0\ mL)} + y \approx 0.033\ M$$

$$[HC_2H_3O_2] = \frac{(0.050\ M)(50.0\ mL)}{(75.0\ mL)} - y \approx 0.033\ M \qquad \frac{y(0.033)}{0.033} = 1.8 \times 10^{-5} = y$$

$$[H_3O^+] = 1.8 \times 10^{-5}\ M \quad \text{thus,} \quad pH = 4.74$$

437. $K_a = 1.80 \times 10^{-5}$

$$HC_2H_3O_2 + H_2O \rightleftharpoons C_2H_3O_2^- + H_3O^+$$

	$HC_2H_3O_2$	$C_2H_3O_2^-$	H_3O^+
Before reaction with NaOH:	0.150	0.200	0.000
After reaction with NaOH:	0.050	0.300	0.000
Change due to equilibrium:	$-x$	$+x$	$+x$
Equilibrium concentrations:	$0.050 - x$	$0.300 + x$	x
	≈ 0.050	≈ 0.300	

$$K_a = \frac{(0.300)(x)}{0.050} = 1.80 \times 10^{-5} \qquad x = 3.0 \times 10^{-6} \quad \text{thus,} \quad pH = 5.52$$

438. The initial hydronium ion concentration of both solutions is 1.8×10^{-5} M. (The reader should verify this statement.)

(A) The 1.8×10^{-5} mol of HCl in the solution will react completely with 1.8×10^{-5} mol of the added NaOH, leaving the excess base in solution.

Excess base $= 0.010$ mol $- (1.8 \times 10^{-5})$ mol $= 0.010$ mol $[OH^-] = \dfrac{0.010 \text{ mol}}{1.0 \text{ L}} = 0.010$ M

The hydroxide ion concentration would be virtually 0.010 M, and the hydronium ion concentration is calculated to be 1.0×10^{-12} M.

(B) The hydroxide ion of the NaOH would react with the 1.8×10^{-5} mol of H_3O^+ present. But as that reaction takes place, more acetic acid ionizes, since removal of the hydronium ion is a stress that shifts the weak acid ionization equilibrium to the right. More and more weak acid dissociates until all the added hydroxide has been used up. Acetate ion is produced in the process. Thus the solution will have a lower acetic acid concentration and a higher acetate ion concentration than it had originally. The solution will be the same as if it had been prepared originally from 0.110 mol of sodium acetate and 0.090 mol of acetic acid in enough water to make 1.00 L of solution.

$$H_2O \quad + \quad HC_2H_3O_2 \quad \rightleftharpoons \quad C_2H_3O_2^- \quad + \quad H_3O^+$$

Before reaction with NaOH:	0.100	0.100	0.000
After reaction with NaOH:	0.090	0.110	0.000
Change due to equilibrium:	$-x$	$+x$	$+x$
Equilibrium concentrations:	$0.090 - x$	$0.110 + x$	x
	≈ 0.090	≈ 0.110	

$$K_a = \frac{x(0.110)}{(0.090)} = 1.80 \times 10^{-5} \quad \text{thus,} \quad x = 1.5 \times 10^{-5} = [H_3O^+]$$

The hydronium ion concentration has been reduced to only five-sixths of its original value by the addition of about 500 times as much OH^- as H_3O^+ originally present. In the buffer solution, the hydronium ion concentration remains relatively constant. Added to the unbuffered solution of HCl, the same quantity of base caused an 18 million-fold change in hydronium ion concentration.

Chapter 21: Heterogeneous and Other Equilibria

439. $AgCl(s) \rightleftharpoons Ag^+ + Cl^-$

$[Ag^+] = [Cl^-] = 1.3 \times 10^{-5}$ $K_{sp} = [Ag^+][Cl^-]$

$K_{sp} = (1.3 \times 10^{-5})(1.3 \times 10^{-5}) = 1.7 \times 10^{-10}$

440. $Mg(OH)_2 \rightleftharpoons Mg^{2+} + 2\,OH^-$ $K_{sp} = [Mg^{2+}][OH^-]^2$

Let $[Mg^{2+}] = x$; then $[OH^-] = 2x$, and $K_{sp} = (x)(2x)^2 = 4x^3 = 7.1 \times 10^{-12}$. Note that $2x$ (not x) is *the* concentration of OH^-, and in this example $[OH^-]$ is twice the concentration of magnesium ion. According to the expression for K_{sp}, it is the hydroxide ion concentration, in this case $2x$, which must be squared.

$$x = 1.2 \times 10^{-4} = [Mg^{2+}] \qquad 2x = 2.4 \times 10^{-4} = [OH^-]$$
$$\text{Check: } K_{sp} = (1.2 \times 10^{-4})(2.4 \times 10^{-4})^2 = 6.9 \times 10^{-12}$$

441. (A) $K_{sp} = [Cu^+][I^-] = 5 \times 10^{-12}$ $[I^-] = 0.010$ M $[Cu^+] = 5 \times 10^{-10}$ M

(B) $A_2X_3 \rightarrow 2A^{3+} + 3X^{2-}$ $K_{sp} = [A^{3+}]^2[X^{2-}]^3 = 1.1 \times 10^{-23}$

        If the solubility of A_2X_3 is y, then $[A^{3+}] = 2y$ and $[X^{2-}] = 3y$.

$(2y)^2(3y)^3 = 108y^5 = 1.1 \times 10^{-23}$ $y^5 = 1.0 \times 10^{-25}$ $y = 1.0 \times 10^{-5}$ mol/L

442. $AgCl \rightleftharpoons Ag^+ + Cl^-$

     Let $[Cl^-] = x$;   then $[Ag^+] = 0.20 + x \approx 0.20$.

                 $0.20x = 1.8 \times 10^{-10}$ (Data from Table 21.1)

                   $x = 9.0 \times 10^{-10}$ M

443. See Questions 439 and 440 for the equations.

     The K_{sp} for AgCl is greater than that for $Mg(OH)_2$, but $Mg(OH)_2$ is more soluble. The apparent anomaly results from the fact that the K_{sp} expressions are not comparable—the $Mg(OH)_2$ expression has a square term in it.

444. $Fe(OH)_3 \rightarrow Fe^{3+} + 3OH^-$

     $K_{sp} = [Fe^{3+}][OH^-]^3 = 1.6 \times 10^{-39}$ $[Fe^{3+}] = \dfrac{1.6 \times 10^{-39}}{(1.0 \times 10^{-8})^3} = 1.6 \times 10^{-15}$ M

445. $[Ag^+][Cl^-] = K_{sp}$ $(4.0 \times 10^{-3})[Cl^-] = 1.8 \times 10^{-10}$ $[Cl^-] = 4.5 \times 10^{-8}$

     Hence, a $[Cl^-]$ of 4.5×10^{-8} M must be exceeded before AgCl precipitates. This question differs from the previous ones in that the two ions forming the precipitate are furnished to the solution independently. This represents a typical analytical situation, in which some soluble chloride is added to precipitate silver ion present in a solution.

446. Assuming that no precipitate forms and the final volume is 200 mL, the lead and chloride ion concentrations would be 0.0500 and 0.150 M, respectively,

$$PbCl_2 \rightleftharpoons Pb^{2+} + 2Cl^- \quad K_{sp} = [Pb^{2+}][Cl^-]^2 = 1.8 \times 10^{-5}$$

     But if there were no precipitation, $[Pb^{2+}][Cl^-]^2 = (0.0500)(0.150)^2 = 1.1 \times 10^{-3}$. Since this product is greater than the value of K_{sp}, the concentrations of ions exceed that required for equilibrium, and precipitation will occur.

447. (A) $Mg(OH)_2 \rightleftharpoons Mg^{2+} + 2OH^-$ $K_{sp} = [Mg^{2+}][OH^-]^2 = 7.1 \times 10^{-12}$

     Any OH^- concentration higher than that contained in a saturated solution would cause precipitation. Hence, the solution must be at the point of attaining equilibrium, and the concentrations of ions in solution must be no greater than those required to satisfy the solubility product constant. In this solution, $[Mg^{2+}] = 0.10$ M.

$$[Mg^{2+}][OH^-]^2 = 7.1 \times 10^{-12}$$

$$[OH^-]^2 = \frac{7.1 \times 10^{-12}}{0.10} = 7.1 \times 10^{-11} \quad [OH^-] = 8.4 \times 10^{-6} \quad pOH = 5.07 \quad pH = 8.93$$

(B) The 10.0 mmol Mg^{2+} added to 20.0 mmol OH^- will yield 10.0 mmol $Mg(OH)_2$. This $Mg(OH)_2$ dissolves according to K_{sp} values.

$$Mg(OH)_2 \rightarrow Mg^{2+} + 2\,OH^- \qquad K_{sp} = [Mg^{2+}][OH^-]^2 = 7.1 \times 10^{-12}$$

At equilibrium, let $[Mg^{2+}] = x$; then $[OH^-] = 2x$.

$$K_{sp} = x(2x)^2 = 4x^3 = 7.1 \times 10^{-12} \qquad x = 1.2 \times 10^{-4} \qquad 2x = 2.4 \times 10^{-4} = [OH^-]$$

448. The two solubilities are not independent of each other because there is a common ion, F^-. We will first assume that most of the F^- in the saturated solution is contributed by the SrF_2, since its K_{sp} is larger than that of CaF_2. We proceed to solve for the solubility of SrF_2 as if the CaF_2 were not present.

If the solubility of SrF_2 is x mol/L, $\quad x = [Sr^{2+}]$ and $2x = [F^-]$. Then

$$(x)(2x)^2 = 4x^3 = K_{sp} = 2.9 \times 10^{-9} \quad \text{or} \quad x = 9.0 \times 10^{-4}$$

The CaF_2 solubility will have to adapt to the concentration of F^- set by the SrF_2 solubility.

$$[Ca^{2+}] = \frac{K_{sp}}{[F^-]^2} = \frac{3.9 \times 10^{-11}}{(2 \times 9.0 \times 10^{-4})^2} = 1.2 \times 10^{-5} \quad \begin{array}{l}\text{(i.e., the solubility of } CaF_2 \text{ is}\\ 1.2 \times 10^{-5} \text{ mol/L)}\end{array}$$

Check of assumption: The amount of F^- contributed by the solubility of CaF_2 is twice the concentration of Ca^{2+}, or 2.4×10^{-5} mol/L. This is indeed small compared with the amount contributed by SrF_2:

$$2 \times 9.0 \times 10^{-4} = 1.8 \times 10^{-3} \text{ mol/L.}$$

449. **(A)** Through **(E)** involve $2\,H_2O \rightarrow H_3O^+ + OH^-$. In addition, **(C)** through **(E)** have the following equilibria:

(C) $HC_2H_3O_2 + H_2O \rightleftharpoons H_3O^+ + C_2H_3O_2^-$

(D) $CuS \rightleftharpoons Cu^{2+} + S^{2-} \qquad S^{2-} + H_2O \rightleftharpoons HS^- + OH^- \qquad HS^- + H_2O \rightleftharpoons H_2S + OH^-$

(E) $NH_3 + H_2O \rightleftharpoons NH_4^+ + OH^- \qquad Mg(OH)_2 \rightleftharpoons Mg^{2+} + 2\,OH^-$

450. We first calculate the S^{2-} concentration:

$$\begin{array}{l} H_2S + H_2O \rightleftharpoons H_3O^+ + HS^- \\ HS^- + H_2O \rightleftharpoons H_3O^+ + S^{2-} \end{array} \qquad K_{12} = \frac{[H_3O^+]^2[S^{2-}]}{[H_2S]} = 9.5 \times 10^{-27}$$

In 0.10 M H_2S and 0.15 M H_3O^+, the concentrations are effectively $[H_3O^+] = 0.15$ and $[H_2S] = 0.10$.

$$K_{12} = \frac{(0.15)^2[S^{2-}]}{(0.10)} = 9.5 \times 10^{-27} \qquad [S^{2-}] = \frac{(9.5 \times 10^{-27})(0.10)}{0.0225} = 4 \times 10^{-26}$$

$$Ag_2S \rightleftharpoons 2Ag^+ + S^{2-} \qquad K_{sp} = [Ag^+]^2[S^{2-}] = 6.3 \times 10^{-50}$$

$$[Ag^+]^2 = \frac{6.3 \times 10^{-50}}{(4 \times 10^{-26})} = 1.6 \times 10^{-24} \quad \text{hence} \quad [Ag^+] = 1.3 \times 10^{-12} \text{ M}$$

451. Not all of the sulfide that dissolves remains as S^{2-}; most of it hydrolyzes:

$$S^{2-} + H_2O \rightleftharpoons HS^- + OH^-$$

452. Initially, assume complete precipitation. $Ag^+ + HCN \rightarrow AgCN + H^+$. Since the solutions were diluted 1:1, $[H^+] = 0.0100$ M. Now consider the equilibria:

$$AgCN \rightleftharpoons Ag^+ + CN^- \qquad K_{sp} = 2.2 \times 10^{-16} = [Ag^+][CN^-]$$

$$HCN \rightleftharpoons H^+ + CN^- \qquad K_a = 6.2 \times 10^{-10} = \frac{[H^+][CN^-]}{[HCN]}$$

Since every dissolved CN^- that hydrolyzes yields one HCN, $[Ag^+] = [CN^-] + [HCN]$.

$$Ag^+ = \frac{2.2 \times 10^{-16}}{[CN^-]} = [CN^-] + \frac{[CN^-](0.0100)}{6.2 \times 10^{-10}} = [CN^-]\left(1 + \frac{0.0100}{6.2 \times 10^{-10}}\right)$$

The 1 in the last term is negligible:

$$[CN^-]^2 = \frac{(2.2 \times 10^{-16})(6.2 \times 10^{-10})}{0.0100} = 1.4 \times 10^{-23} \qquad [CN^-] = 3.7 \times 10^{-12}$$

$$[Ag^+] = \frac{K_{sp}}{[CN^-]} = \frac{2.2 \times 10^{-16}}{3.7 \times 10^{-12}} = 5.9 \times 10^{-5}$$

453. The precipitation reaction yields H_3O^+, and the H_2S will become saturated (0.10 M) after the precipitation reaction.

$$H_2S + 2H_2O + Cu^{2+} \rightarrow CuS + 2H_3O^+ \qquad (0.15 \text{ M Cu})\left(\frac{2 \text{ mol } H_3O^+}{\text{mol } Cu^{2+}}\right) = 0.30 \text{ M } H_3O^+$$

See the equations in Question 450.

$$K_{12} = \frac{(0.30)^2[S^{2-}]}{0.10} = 9.5 \times 10^{-27} \qquad [S^{2-}] = 1 \times 10^{-26} \quad \text{(one significant figure)}$$

$$[Cu^{2+}] = \frac{K_{sp}}{[S^{2-}]} = \frac{8.5 \times 10^{-36}}{1 \times 10^{-26}} = 8 \times 10^{-10} \text{ M}$$

Chapter 22: Electrochemistry

454. volt, ampere, joule, watt, current, potential, faraday

455. **(A)** watt **(B)** coulomb **(C)** ohm **(D)** coulomb **(E)** ampere **(F)** watt **(G)** volt **(H)** ohm

456. Energy is equal to the product of potential times charge or potential times current times time.

$$(115 \text{ V})(1.00 \text{ A})(100 \text{ s}) = 11,500 \text{ J} = 11.5 \text{ kJ}$$

457. $\left(\dfrac{96,500 \text{ C}}{\text{mol } e^-}\right)\left(\dfrac{1 \text{ mol } e^-}{6.02 \times 10^{23} e^-}\right) = 1.60 \times 10^{-19} \text{ C/}e^-$

458. $Cu^{2+} + 2e^- \rightarrow Cu$ or $2 Cl^- \rightarrow Cl_2 + 2e^-$

Two moles of electronic charge ($2\mathscr{F}$) is required per mol $CuCl_2$.

459. The electrode reactions are

Anode: $Cu \rightarrow Cu^{2+} + 2e^-$

Cathode: $Cu^{2+} + 2e^- \rightarrow Cu$

$$(10.0 \text{ A})(3600 \text{ s})\left(\dfrac{1 \text{ C}}{1 \text{ A} \cdot \text{s}}\right)\left(\dfrac{1 \text{ mol } e^-}{96,500 \text{ C}}\right) = 0.373 \text{ mol } e^-$$

Since it takes 2 mol electrons to react with or produce each mole of copper metal, the number of moles of copper dissolved or deposited is

$$(0.373 \text{ mol } e^-)\left(\dfrac{1 \text{ mol Cu}}{2 \text{ mol } e^-}\right) = 0.186 \text{ mol Cu}$$

Thus, 0.186 mol copper is dissolved from the anode, 0.186 mol copper is deposited onto the cathode, and the original copper(II) ion concentration of the solution remains unchanged.

460. (A) $Hg^{2+} + 2e^- \rightarrow Hg$ $(19,300 \text{ C})\left(\dfrac{1 \text{ mol } e^-}{96,500 \text{ C}}\right)\left(\dfrac{1 \text{ mol Hg}}{2 \text{ mol } e^-}\right)\left(\dfrac{201 \text{ g Hg}}{\text{mol Hg}}\right) = 20.1 \text{ g}$

(B) $Hg_2^{2+} + 2e^- \rightarrow 2Hg$ $(19,300 \text{ C})\left(\dfrac{1 \text{ mol } e^-}{96,500 \text{ C}}\right)\left(\dfrac{2 \text{ mol Hg}}{2 \text{ mol } e^-}\right)\left(\dfrac{201 \text{ g Hg}}{\text{mol Hg}}\right) = 40.2 \text{ g}$

461. $(100 \text{ g Ni})\left(\dfrac{1 \text{ mol Ni}}{58.7 \text{ g Ni}}\right)\left(\dfrac{2 \text{ mol } e^-}{\text{mol Ni}}\right)\left(\dfrac{96,500 \text{ C}}{\text{mol } e^-}\right) = 3.29 \times 10^5 \text{C}$

$(200 \text{ min})(60 \text{ s/min}) = 1.20 \times 10^4 \text{s}$ $\dfrac{3.29 \times 10^5 \text{ C}}{1.20 \times 10^4 \text{ s}} = 27.4 \text{ A}$

462. $(107.9 \text{ g Ag})\left(\dfrac{1 \text{ mol Ag}}{107.9 \text{ g}}\right)\left(\dfrac{1 \text{ mol } e^-}{\text{mol Ag}}\right) = 1.000 \text{ mol } e^-$

$(1.000 \text{ mol } e^-) = (96,500 \text{ C})\left(\dfrac{1 \text{ electron}}{1.60 \times 10^{-19} \text{ C}}\right) = 6.03 \times 10^{23} e^-$

Avogadro's number, 6.02×10^{23}, is equal to the number of electrons per mole of electrons.

463. (A) The number of moles of tin and the number of moles of electrons are determined.

$$(5.00 \text{ h})\left(\frac{3600 \text{ s}}{\text{h}}\right)\left(\frac{2.00 \text{ C}}{\text{s}}\right)\left(\frac{1 \text{ mol } e^-}{96{,}500 \text{ C}}\right) = 0.373 \text{ mol } e^-$$

$$(22.2 \text{ g Sn})\left(\frac{1 \text{ mol Sn}}{118.69 \text{ g Sn}}\right) = 0.187 \text{ mol Sn}$$

$$\frac{0.373 \text{ mol } e^-}{0.187 \text{ mol Sn}} = \frac{2 \text{ mol } e^-}{\text{mol Sn}} \qquad \text{The oxidation state is } +2.$$

(B)
$$(600 \text{ s})(0.200 \text{ A})\left(\frac{1 \text{ mol } e^-}{96{,}500 \text{ C}}\right) = 1.24 \times 10^{-3} \text{ mol } e^-$$

$$Cl^- \rightarrow \tfrac{1}{2} Cl_2 + e^- \qquad H_2O + e^- \rightarrow \tfrac{1}{2} H_2 + OH^-$$

The oxidation does not affect the OH^- concentration. The reduction produces

$$(1.24 \times 10^{-3} \text{ mol } e^-)\left(\frac{1 \text{ mol OH}^-}{\text{mol } e^-}\right) = 1.24 \times 10^{-3} \text{ mol OH}^- = 1.24 \text{ mmol OH}^-$$

$$\frac{1.24 \text{ mmol OH}^-}{50.00 \text{ mL}} = 0.0248 \text{ M OH}^-$$

464. (A) The equations for the half-reactions, with the corresponding standard potentials, are

$$Fe^{3+} + e^- \rightarrow Fe^{2+} \qquad \varepsilon^\circ = +0.77 \text{ V}$$
$$Zn \rightarrow Zn^{2+} + 2e^- \qquad \varepsilon^\circ = +0.76 \text{ V}$$

(B) To obtain the equation for the cell reaction, the equation for the iron(III)/iron(II) half-reaction must be multiplied by 2 (to obtain 2 mol electrons) and combined with the zinc/zinc ion half-reaction. The value of the half-cell potential is *not* multiplied by 2, however, because potential does *not* depend on the *quantity* of substance undergoing reaction. Addition of the resulting equations and the two potentials results in the following cell reaction:
$$Zn + 2 Fe^{3+} \rightarrow 2 Fe^{2+} + Zn^{2+} \qquad \varepsilon^\circ = 1.53 \text{ V}$$

The sign of ε_{cell} determines the direction of reaction. The positive sign means that the reaction is spontaneous as written. (A negative sign would mean that the reverse reaction is spontaneous.)

465. The desired reaction is $2 Ag + 2 H^+(1.0 \text{ M}) \rightarrow 2 Ag^+(1.0 \text{ M}) + H_2(1 \text{ atm})$. Combining the two half-reactions gives

$$2 Ag \rightarrow 2 Ag^+ + 2e^- \qquad \varepsilon^\circ = -0.80 \text{ V}$$
$$\underline{2 H^+ + 2e^- \rightarrow H_2 \qquad\qquad \varepsilon^\circ = \;\;\, 0.00 \text{ V}}$$
$$2 Ag + 2 H^+ \rightarrow H_2 + 2 Ag^+ \qquad \varepsilon^\circ_{cell} = -0.80 \text{ V}$$

The large negative value of the standard cell potential indicates that H^+ solution does not oxidize silver metal under these conditions.

466. Zinc ion would not undergo any reaction in the copper half-cell. Copper(II) ion, however, would be reduced to copper on contact with the zinc electrode in the Zn^{2+}/Zn half-cell. During discharge, zinc ion might get into the Cu^{2+}/Cu half-cell, which is alright. During recharge, copper(II) ion might get into the Zn^{2+}/Zn half-cell, where it would plate out on the electrode and destroy the cell.

467. (A) $2e^- + Sn^{2+} \rightarrow Sn \qquad \varepsilon^\circ = -0.14 \text{ V}$

$$Sn^{2+} \rightarrow Sn^{4+} + 2e^- \quad \varepsilon^\circ = -0.13 \text{ V}$$

$$2Sn^{2+} \rightarrow Sn + Sn^{4+} \quad \varepsilon^\circ = -0.27 \text{ V}$$

The disproportionation reaction is not spontaneous, because the ε° value is negative. Hence, tin(II) is stable.

    (B) Water is less difficult to reduce ($\varepsilon^\circ = -0.414$ V, with $[H_3O^+] = 1.0 \times 10^{-7}$ M) than aluminum ion ($\varepsilon^\circ = -1.66$ V). Na_3AlF_6 dissolves Al_2O_3 but is not more easily reduced than aluminum ion. Al_2O_3 cannot be melted economically; it is a refractory substance, used for furnace linings, with a melting point of 2045°C.

    (C) $e^- + Co^{3+} \rightarrow Co^{2+} \quad \varepsilon^\circ = 1.82$ V   This potential is much more than that necessary to oxidize water.

468. (A) Fe ($Fe^{3+} \rightarrow Fe^{2+} \rightarrow Fe$) and O_2 (H_2O is oxidized more easily than SO_4^{2-})   **(B)** H_2 and AgCl (H_2O is reduced more easily than Li^+; as Ag is oxidized in the presence of Cl^- ion, AgCl forms).   **(C)** Na and F_2.

469. $(1.00 \text{ h})\left(\dfrac{3600 \text{ s}}{\text{h}}\right)\left(\dfrac{4.00 \text{ C}}{\text{s}}\right)\left(\dfrac{1 \text{ mol } e^-}{96,500 \text{ C}}\right) = 0.149 \text{ mol } e^-$

$$\left(\dfrac{0.100 \text{ mol } Fe^{3+}}{L}\right)(1.00 \text{ L}) = 0.100 \text{ mol } Fe^{3+}$$

$$Fe^{3+} + e^- \rightarrow Fe^{2+}$$

Since 0.100 mol e^- is required to reduce all the Fe^{3+} to Fe^{2+}, 0.049 mol e^- will be left to reduce the Fe^{2+} to Fe.

$$Fe^{2+} + 2e^- \rightarrow Fe \qquad\qquad (0.049 \text{ mol } e^-)\left(\dfrac{1 \text{ mol Fe}}{2 \text{ mol } e^-}\right) = 0.025 \text{ mol Fe}$$

470. $\qquad\qquad Zn^{2+} + 2e^- \rightarrow Zn \qquad \varepsilon^\circ = -0.763$ V

$$\varepsilon = \varepsilon^\circ - \left(\dfrac{0.0592}{n}\right)\log\left(\dfrac{a(Zn)}{[Zn^{2+}]}\right)$$

or $\qquad\qquad \varepsilon = \varepsilon^\circ - \left(\dfrac{0.0257}{n}\right)\ln\left(\dfrac{a(Zn)}{[Zn^{2+}]}\right)$

    where ln is the natural logarithm (LN key on a simple scientific calculator) and 0.0257 is the value of $RT/\mathcal{F}$ at 25°C. (It is simpler to memorize one of these numbers than to calculate it for each question.)

The activity of pure zinc metal, $a(Zn)$, is 1.00, and the Zn^{2+} concentration is 0.0100 M.

$$\varepsilon = (-0.763) - \left(\frac{0.0592}{2}\right)\left(\log\frac{1.00}{0.0100}\right)$$

$$= (-0.763) - \left(\frac{0.0592}{2}\right)[\log(100)]$$

$$= -0.763 - 0.0592 = -0.822\,V$$

471. $NO_3^- + 4H^+ + 3e^- \rightarrow NO + 2H_2O$

Set **(A)**. By definition $\varepsilon°$ refers to a reaction in which all reactants and products are at unit activity (1 M solutes, 1 atm gases, pure solids or liquids).

472. (A) $Pb \rightarrow Pb^{2+} + 2e^-$ $\varepsilon° = 0.13\,V$

$\quad\quad\quad\quad Sn^{2+} + 2e^- \rightarrow Sn$ $\varepsilon° = -0.14\,V$

$\quad\quad\quad\quad Sn^{2+} + Pb \rightarrow Pb^{2+} + Sn$ $\varepsilon° = -0.01\,V$

Sn and Pb are each at unity activity, since each is a pure metal.

$$\varepsilon = \varepsilon° - \left(\frac{0.0592}{2}\right)\left(\log\frac{[Pb^{2+}]}{[Sn^{2+}]}\right) = -0.01 - (0.0296)\left(\log\frac{[Pb^{2+}]}{[Sn^{2+}]}\right) = 0$$

$$\log\frac{[Pb^{2+}]}{[Sn^{2+}]} = \frac{0.01}{-0.0296} = -0.3 \quad\quad\quad \frac{[Pb^{2+}]}{[Sn^{2+}]} = 0.5$$

(B) $\varepsilon = 0$ is for a cell in which the concentration ratio term is equal to $\varepsilon°$, a system at equilibrium. $\varepsilon° = 0$ indicates a concentration cell, a cell in which one half-reaction is the reverse of the other.

473. We can use either the log or ln form of the Nernst equation.

$$Fe^{3+} + e^- \rightarrow Fe^{2+}$$

$$\varepsilon = \varepsilon° - \left(\frac{0.0257}{1}\right)\left(\ln\frac{[Fe^{2+}]}{[Fe^{3+}]}\right) = 0.771 - 0.0257\left(\ln\frac{2.0}{0.020}\right) = 0.771 - (0.0257)(\ln 100)$$

$$= 0.653\,V$$

474. $Hg_2Cl_2 + 2e^- \rightarrow 2Hg + 2Cl^-$ $\quad\quad\quad\quad \varepsilon° = 0.270\,V$

$\quad\quad\quad\quad 2Cl^- \rightarrow Cl_2 + 2e^-$ $\quad\quad\quad\quad\quad\quad \varepsilon° = -1.36\,V$

$\quad\quad\quad\quad Hg_2Cl_2(s) \rightarrow 2Hg(l) + Cl_2(0.80\,atm)$ $\quad \varepsilon° = -1.09\,V$

$$\varepsilon = \varepsilon° - \left(\frac{0.0592}{2}\right)[\log P(Cl_2)] = (-1.09) - \left(\frac{0.0592}{2}\right)(\log 0.80) = -1.09\,V$$

Note that neither the solid Hg_2Cl_2 nor the liquid Hg appears in the Nernst equation.

475. (A)

$$Zn \rightarrow Zn^{2+}(1.0 \text{ M}) + 2e^- \quad \varepsilon° = 0.763 \text{ V}$$
$$Zn^{2+}(0.15 \text{ M}) + 2e^- \rightarrow Zn \quad \varepsilon° = -0.763 \text{ V}$$
$$Zn^{2+}(0.15 \text{ M}) \rightarrow Zn^{2+}(1.0 \text{ M}) \quad \varepsilon° = 0.000 \text{ V}$$

$$\varepsilon = 0.000 - \left(\frac{0.0592}{2}\right)\left(\log\frac{1.0}{0.15}\right) = -(0.0296)(0.824) = -0.0244 \text{ V}$$

As the cell discharges (the reaction proceeds to the left, since the value of ε is negative), the 1.0 M zinc ion is used up and the 0.15 M zinc ion is produced. Thus, the two solutions approach each other in concentration (just as they would if the solutions were mixed directly).

(B)

$$2H^+ + 2e^- \rightarrow H_2 \quad \varepsilon° = 0.000 \text{ V}$$

$$\varepsilon = -0.414 = (0.000) - \left(\frac{0.0257}{2}\right)\left(\ln\frac{1}{[H^+]^2}\right) = (0.0257)(\ln [H^+])$$

$$\ln [H^+] = -\frac{0.414}{0.0257} = -16.1 \qquad [H^+] = 1.01 \times 10^{-7} \text{ M}$$

476. $Cu^{2+} + 2e^- \rightarrow Cu \qquad \varepsilon° = 0.34 \text{ V}$

$$\varepsilon = 0.34 - \left(\frac{0.0592}{2}\right)\left(\log\frac{1}{[Cu^{2+}]}\right) = 0.34 - (0.0296)(9.00) = 0.34 - 0.266 = 0.07 \text{ V}$$

Electrolysis at 0.07 V would leave 1.00×10^{-9} M Cu^{2+} without reducing the Ni^{2+} concentration. [At 0.07 V, the concentration of nickel(II) ion could theoretically be as high as 6×10^{10} M. The calculation follows.]

$$Ni^{2+} + 2e^- \rightarrow Ni \quad \varepsilon° = -0.25 \text{ V}$$

$$0.07 = -0.25 - \left(\frac{0.0592}{2}\right)\left(\log\frac{1}{[Ni^{2+}]}\right)$$

$$0.32 = (0.0296)(\log[Ni^{2+}]) \qquad [Ni^{2+}] = 6 \times 10^{10} \text{ M}$$

477. Magnesium acts, by cathodic protection, to prevent oxidation of the steel by transferring an excess of electrons to the steel.

478. (A) The reagents are solids or concentrated solutes. **(B)** Discharge is accompanied by conversion of (dense) H_2SO_4 to (less dense) H_2O. **(C)** The oxidizing and reducing agents, in both discharge and recharge, are solids. Thus all the reagents can be placed in the same vessel. **(D)** The solid nature of each oxidizing and reducing agent prevents direct contact no matter which way the reactions are run. [Or the same answer as **(C)**.]

479. $\varepsilon = 0$ at equilibrium. At equilibrium, the concentration ratio equals K.

480. (A) $\varepsilon°$(cell) is obtained as follows:

$$Zn \rightarrow Zn^{2+} + 2e^- \qquad \varepsilon° = 0.76 \text{ V}$$
$$\underline{Cu^{2+} + 2e^- \rightarrow Cu \qquad \varepsilon° = 0.34 \text{ V}}$$
$$Zn + Cu^{2+} \rightarrow Zn^{2+} + Cu \qquad \varepsilon°(\text{cell}) = 1.10 \text{ V}$$

Since 2 mol electrons is associated with 1 mol of chemical reaction,

$$\Delta G° = -n\mathscr{F}\varepsilon°(\text{cell}) = -(2 \text{ mol } e^-)\left(\frac{96,500 \text{ C}}{\text{mol } e^-}\right)(1.10 \text{ V}) = -212,000 \text{ J} = -212 \text{ kJ}$$

(B) Maintaining the concentrations at 1.00 M allows use of $\varepsilon°$ values to get ΔG. Any constant concentrations could have been used. Using Table 22.1,

$$Ag^+ + e^- \rightarrow Ag \qquad \varepsilon° = 0.80 \text{ V}$$
$$\underline{Cu^{2+} + 2e^- \rightarrow Cu \qquad \varepsilon° = 0.34 \text{ V}}$$
$$Cu + 2Ag^+ \rightarrow 2Ag + Cu^{2+} \qquad \varepsilon° = 0.46 \text{ V}$$

$$\Delta G° = -\varepsilon°n\mathscr{F} = -(0.46 \text{ V})(2 \text{ mol } e^-)\left(\frac{96,500 \text{ C}}{\text{mol } e^-}\right) = -89,000 \text{ J} = -89 \text{ kJ}$$

481. $\varepsilon = \varepsilon° - \left(\dfrac{RT}{n\mathscr{F}}\right)(\ln Q)$ where ln means natural logarithm and Q represents the ratio

of concentrations.

(We have reversed the derivation of the Nernst equation.)

$$\varepsilon = \frac{-\Delta G}{n\mathscr{F}} = \frac{-\Delta G°}{n\mathscr{F}} - \left(\frac{RT}{n\mathscr{F}}\right)(\ln Q)$$

$$\Delta G = \Delta G° + (RT)(\ln Q) = \Delta G° + (2.303RT)(\log Q)$$

Chapter 23: Nuclear and Radiochemistry

482. (A) Isotopes have the same atomic numbers: $^{40}_{21}Sc$ and $^{42}_{21}Sc$. **(B)** Isobars have the same mass numbers: $^{40}_{18}Ar$ and $^{40}_{21}Sc$.

483. Since

$$1 \text{ u} = \frac{1}{6.02 \times 10^{23}} \text{ g} = \frac{1}{6.02 \times 10^{26}} \text{ kg} \quad \text{and} \quad 1 \text{ J} = 1 \text{ kg} \cdot \text{m}^2/\text{s}^2$$

$$E = mc^2 = \left(\frac{1 \text{ kg}}{6.02 \times 10^{26}}\right)(3.00 \times 10^8 \text{ m/s})^2 = 1.49 \times 10^{-10} \text{ J}$$

But

$$1 \text{ MeV} = 10^6 \text{ eV} = (10^6)(1.60 \times 10^{-19} \text{ C})(1.00 \text{ V}) = 1.60 \times 10^{-13} \text{ J}$$

and so

$$E = (1.49 \times 10^{-10} \text{ J})\left(\frac{1 \text{ MeV}}{1.60 \times 10^{-13} \text{ J}}\right) = 931 \text{ MeV}$$

(A more precise value is 931.5 MeV/u.)

484. No. Mass numbers reflect actual masses only to the nearest u. The precise masses are 1.0072765 u for the proton and 1.0086650 u for the neutron.

485. Protons: 94, from the periodic table. Electrons: 94, since the atom is uncharged. Neutrons: $239 - 94 = 145$.

486. Since each element has only one naturally occurring isotope, the mass number is the integer closest to the atomic mass. $^{19}_{9}F$, $^{75}_{33}As$

487. **(A)** The sum of the subscripts on the left is $7 + 2 = 9$. The subscript of the first product on the right is 8. Hence the second product on the right must have a subscript (nuclear charge) of 1. The sum of the superscripts on the left is $14 + 4 = 18$. The superscript of the first product on the right is 17. Hence the second product on the right must have a superscript (mass number) of 1. The particle with nuclear charge 1 and mass number 1 is the proton, $^{1}_{1}H$.

The other parts are solved in a similar manner.

    (B) $^{1}_{0}n$ **(C)** $^{6}_{3}Li$ **(D)** $^{0}_{-1}\beta$ **(E)** $^{0}_{-1}e$ **(F)** $^{1}_{1}p$

488. In emitting three alpha particles, six protons (as well as six neutrons) are lost from the nucleus. The atomic number is reduced by 6, and the resulting element is in group IVA of the preceding period.

489. $^{208}_{82}Pb$ (It is a member of the $4n$ series, as is ^{232}Th.)

490. The subscripts already balance, so the subscript of the unknown particle(s) must be zero. The superscripts on the right total 3 less than those on the left. There must be three neutrons $\left(^{1}_{0}n\right)$ also produced.

$$^{235}_{92}U + {}^{1}_{0}n \rightarrow {}^{139}_{54}Xe + {}^{94}_{38}Sr + 3\ {}^{1}_{0}n$$

Note that gamma particles $\left(^{0}_{0}\gamma\right)$ might also be produced, but since they have both zero subscript and zero superscript, their presence cannot be deduced by this kind of analysis.

491. **(A)** beta emission   **(B)** alpha emission, positron emission, and electron capture   **(C)** electron capture (The capture of a K shell electron leaves a vacancy, which is filled by a higher-energy electron. This transfer is accompanied by x-ray emission.)

492. 80 y is four half-lives; it will decay to $\left(\frac{1}{2}\right)^{4}(8000\ dis/min) = 500\ dis/min$.

493. $A = \lambda N$ $\dfrac{A}{A_0} = \dfrac{\lambda N}{\lambda N_0} = \dfrac{N}{N_0} = \dfrac{m}{m_0}$, the last equality because of proportionality of mass and number of atoms of a given nuclide.

494. The 90% decay corresponds to 10% (or 0.10) survival.

$$k = -\frac{\ln([A])/([A]_0)}{t} = -\frac{\ln 0.10}{366 \text{ min}} = 6.29 \times 10^{-3} \text{ min}^{-1}$$

Then the half-life can be computed:

$$t_{1/2} = \frac{\ln 2}{k} = \frac{0.693}{6.29 \times 10^{-3} \text{ min}^{-1}} = 110 \text{ min}$$

495. (A) $\left(\ln\dfrac{m}{m_0}\right) = -\left(\dfrac{\ln 2}{t_{1/2}}\right)t = -\left(\dfrac{0.693}{4.50 \times 10^9 \text{ y}}\right)(3.00 \times 10^9 \text{ y}) = -0.462$

$m/m_0 = 0.630$ $\qquad\qquad m = m_0(0.630) = (2.00 \text{ kg})(0.630) = 1.26 \text{ kg left}$

(B) Of the 2.00 kg ^{238}U initially present, 2.00 kg − 1.26 kg = 0.74 kg disintegrated. For every U atom that disintegrates, one Pb atom is produced. The mass ratio is just 206/238. Hence,

$$(0.74 \text{ kg } {}^{238}\text{U})\left(\frac{0.206 \text{ kg } {}^{206}\text{Pb}}{0.238 \text{ kg } {}^{238}\text{U}}\right) = 0.64 \text{ kg } {}^{206}\text{Pb produced}$$

496. (A) Two half-lives, hence 0.25.

(B) $\ln f = -\left(\dfrac{0.693}{5730 \text{ y}}\right)(13,000 \text{ y}) = -1.57$ $\qquad f = 0.208$

497. Difference = m(U atom) − m(He atom) − m(Th atom)

$\qquad\qquad$ = $[m$(U nucleus) + $92m_e] - [m$(He nucleus) + $2m_e] - [m$(Th nucleus) + $90m_e]$

$\qquad\qquad$ = m(U nucleus) − m(He nucleus) − m(Th nucleus)

$\qquad$ (There is a very small difference, due to electronic binding energies, between the rest mass of a neutral atom and the sum of the nuclear and electronic rest masses.)

498. Taking values from Table 23.2,

$\qquad$ mass defect = $2(1.0072765 \text{ u}) + 2(1.0086650 \text{ u}) - 4.001506 \text{ u} = 0.030377 \text{ u}$

To get the $^4_2$He nucleus to come apart into its four nucleons, energy equivalent to 0.0304 u would have to be added.

$$E = (0.0304 \text{ u})(932 \text{ MeV/u}) = 28.3 \text{ MeV}$$

499. %Zn recovered = % ^{62}Zn recovered = $\dfrac{0.0823}{0.100} \times 100 = 82.3\%$

$$0.2000 \text{ g recovered}\left(\frac{100 \text{ g total}}{82.3 \text{ g recovered}}\right) = 0.243 \text{ g total}$$

The mass of the added $^{62}Zn^{2+}$ is negligible; hence, 0.243 g is the mass of the Zn^{2+} in the original sample.

$$0.243 \text{ g} \left(\frac{1 \text{ mol}}{65.37 \text{ g}} \right) = 3.72 \times 10^{-3} \text{ mol} \qquad \frac{3.72 \times 10^{-3} \text{ mol}}{0.05000 \text{ L}} = 0.0744 \text{ M } Zn^{2+}$$

500. $\frac{1}{8} S_8^* + SO_3^{2-} \rightarrow S^*SO_3^{2-} \quad (S_2O_3^{2-})$

$Ba^{2+} + S^*SO_3^{2-} \rightarrow BaS^*SO_3$

$BaS^*SO_3 + 2 H_3O^+ \rightarrow SO_2 + \frac{1}{8} S_8^* + 3 H_2O + Ba^{2+}$

The two sulfur atoms in the thiosulfate ion are not equivalent and do not "exchange" with one another. The structure of the thiosulfate ion is

$$\begin{array}{c} \ddot{\text{:}\ddot{\text{O}}\text{:}} \quad {}^{2-} \\ \text{:}\ddot{\text{O}}\text{:}\,\ddot{\text{S}}\,\text{:}\,\ddot{\text{S}}^*\text{:} \\ \text{:}\ddot{\text{O}}\text{:} \end{array}$$

Table of the Elements

Element	Symbol	Atomic Number	Atomic Mass		Element	Symbol	Atomic Number	Atomic Mass
Actinium	Ac	89	(227)		Dysprosium	Dy	66	162.50
Aluminum	Al	13	26.9815		Einsteinium	Es	99	(254)
Americium	Am	95	(243)		Erbium	Er	68	167.26
Antimony	Sb	51	121.75		Europium	Eu	63	151.96
Argon	Ar	18	39.948		Fermium	Fm	100	(253)
Arsenic	As	33	74.9216		Fluorine	F	9	18.9984
Astatine	At	85	(210)		Francium	Fr	87	(223)
Silver	Ag	47	107.868		Iron	Fe	26	55.847
Gold	Au	79	196.9665		Gadolinium	Gd	64	157.25
Barium	Ba	56	137.34		Gallium	Ga	31	69.72
Berkelium	Bk	97	(249)		Germanium	Ge	32	72.59
Beryllium	Be	4	9.01218		Gold	Au	79	196.9665
Bismuth	Bi	83	208.9806		Hafnium	Hf	72	178.49
Bohrium	Bh	107	(262)		Hassium	Hs	108	(265)
Boron	B	5	10.81		Helium	He	2	4.00260
Bromine	Br	35	79.904		Holmium	Ho	67	164.9303
Cadmium	Cd	48	112.40		Hydrogen	H	1	1.0080
Calcium	Ca	20	40.08		Mercury	Hg	80	200.59
Californium	Cf	98	(251)		Indium	In	49	114.82
Carbon	C	6	12.011		Iodine	I	53	126.9045
Cerium	Ce	58	140.12		Iridium	Ir	77	192.22
Cesium	Cs	55	132.9055		Iron	Fe	26	55.847
Chlorine	Cl	17	35.453		Krypton	Kr	36	83.80
Chromium	Cr	24	51.996		Potassium	K	19	39.102
Cobalt	Co	27	58.9332		Lanthanum	La	57	138.9055
Copper	Cu	29	63.546		Lawrencium	Lr	103	(257)
Curium	Cm	96	(247)		Lead	Pb	82	207.2
Dubnium	Db	105	(260)		Lithium	Li	3	6.941

Element	Symbol	Number	Weight	Element	Symbol	Number	Weight
Lutetium	Lu	71	174.97	Ruthenium	Ru	44	101.07
Magnesium	Mg	12	24.305	Rutherfordium	Rf	104	(257)
Manganese	Mn	25	54.9380	Samarium	Sm	62	150.4
Meitnerium	Mt	109	(266)	Scandium	Sc	21	44.9559
Mendelevium	Md	101	(256)	Scaborgium	Sg	106	(263)
Mercury	Hg	80	200.59	Selenium	Se	34	78.96
Molybdenum	Mo	42	95.94	Silicon	Si	14	28.086
Neodymium	Nd	60	144.24	Silver	Ag	47	107.868
Neon	Ne	10	20.179	Sodium	Na	11	22.9898
Neptunium	Np	93	237.0482	Strontium	Sr	38	87.62
Nickel	Ni	28	58.71	Sulfur	S	16	32.06
Niobium	Nb	41	92.9064	Antimony	Sb	51	121.75
Nitrogen	N	7	14.0067	Tin	Sn	50	118.69
Nobelium	No	102	(254)	Tantalum	Ta	73	180.9479
Sodium	Na	11	22.9898	Technetium	Tc	43	98.9062
Osmium	Os	76	190.2	Tellurium	Te	52	127.60
Oxygen	O	8	15.9994	Terbium	Tb	65	158.9254
Palladium	Pd	46	106.4	Thallium	Tl	81	204.37
Phosphorus	P	15	30.9738	Thorium	Th	90	232.0381
Platinum	Pt	78	195.09	Thulium	Tm	69	168.9342
Plutonium	Pu	94	(242)	Tin	Sn	50	118.69
Polonium	Po	84	(210)	Titanium	Ti	22	47.90
Potassium	K	19	39.102	Tungsten	W	74	183.85
Praseodymium	Pr	59	140.9077	Uranium	U	92	238.029
Promethium	Pm	61	(145)	Vanadium	V	23	50.9414
Protactinium	Pa	91	231.0359	Tungsten	W	74	183.85
Lead	Pb	82	207.2	Xenon	Xe	54	131.30
Radium	Ra	88	226.0254	Ytterbium	Yb	70	173.04
Radon	Rn	86	(222)	Yttrium	Y	39	88.9059
Rhenium	Re	75	186.2	Zinc	Zn	30	65.37
Rhodium	Rh	45	102.9055	Zirconium	Zr	40	91.22
Rubidium	Rb	37	85.4678				

Periodic Table of the Elements

Key:

1
H
1.0080

Atomic number
Symbol
Atomic mass

Classical group numbers	IA	IIA	IIIB	IVB	VB	VIB	VIIB	VIII	VIII	VIII	IB	IIB	IIIA	IVA	VA	VIA	VIIA	0
Modern group numbers	1	2	3	4	5	6	7	8	9	10	11	12	13	14	15	16	17	18
Periods																		
1	1 **H** 1.0080																	2 **He** 4.00260
2	3 **Li** 6.941	4 **Be** 9.01218											5 **B** 10.81	6 **C** 12.011	7 **N** 14.0067	8 **O** 15.9994	9 **F** 18.9984	10 **Ne** 20.179
3	11 **Na** 22.9898	12 **Mg** 24.305											13 **Al** 26.9815	14 **Si** 28.086	15 **P** 30.9738	16 **S** 32.06	17 **Cl** 35.453	18 **Ar** 39.948
4	19 **K** 39.102	20 **Ca** 40.08	21 **Sc** 44.9559	22 **Ti** 47.90	23 **V** 50.9414	24 **Cr** 51.996	25 **Mn** 54.9380	26 **Fe** 55.847	27 **Co** 58.9332	28 **Ni** 58.71	29 **Cu** 63.546	30 **Zn** 65.37	31 **Ga** 69.72	32 **Ge** 72.59	33 **As** 74.9216	34 **Se** 78.96	35 **Br** 79.904	36 **Kr** 83.80
5	37 **Rb** 85.4678	38 **Sr** 87.62	39 **Y** 88.9059	40 **Zr** 91.22	41 **Nb** 92.9064	42 **Mo** 95.94	43 **Tc** 98.9062	44 **Ru** 101.07	45 **Rh** 102.9055	46 **Pd** 106.4	47 **Ag** 107.868	48 **Cd** 112.40	49 **In** 114.82	50 **Sn** 118.69	51 **Sb** 121.75	52 **Te** 127.60	53 **I** 126.9045	54 **Xe** 131.30
6	55 **Cs** 132.9055	56 **Ba** 137.34	57 **La** 138.9055 •	72 **Hf** 178.49	73 **Ta** 180.9479	74 **W** 183.85	75 **Re** 186.2	76 **Os** 190.2	77 **Ir** 192.22	78 **Pt** 195.09	79 **Au** 196.9665	80 **Hg** 200.59	81 **Tl** 204.37	82 **Pb** 207.2	83 **Bi** 208.9806	84 **Po** (210)	85 **At** (210)	86 **Rn** (222)
7	87 **Fr** (223)	88 **Ra** 226.0254	89 **Ac** (227) †	104 **Rf** (257)	105 **Db** (260)	106 **Sg** (263)	107 **Bh** (262)	108 **Hs** (265)	109 **Mt** (266)	110	111	112	(113)	114	(115)	116	(117)	118

•

58 **Ce** 140.12	59 **Pr** 140.9077	60 **Nd** 144.24	61 **Pm** (145)	62 **Sm** 150.4	63 **Eu** 151.96	64 **Gd** 157.25	65 **Tb** 158.9254	66 **Dy** 162.50	67 **Ho** 164.9303	68 **Er** 167.26	69 **Tm** 168.9342	70 **Yb** 173.04	71 **Lu** 174.97

†

90 **Th** 232.0381	91 **Pa** 231.0359	92 **U** 238.029	93 **Np** 237.0482	94 **Pu** (242)	95 **Am** (243)	96 **Cm** (247)	97 **Bk** (249)	98 **Cf** (251)	99 **Es** (254)	100 **Fm** (253)	101 **Md** (256)	102 **No** (254)	103 **Lr** (257)